# Geomagnetic Disturbances Impacts on Power Systems

# Geomagnetic Disturbances Impacts on Power Systems

## Risk Analysis and Mitigation Strategies

Olga Sokolova
Nikolay Korovkin
Masashi Hayakawa

## CRC Press

Taylor & Francis Group

Boca Raton London New York

CRC Press is an imprint of the
Taylor & Francis Group, an **informa** business

First edition published 2021
by CRC Press
6000 Broken Sound Parkway NW, Suite 300, Boca Raton, FL 33487-2742

and by CRC Press
2 Park Square, Milton Park, Abingdon, Oxon, OX14 4RN

© 2021 Olga Sokolova

CRC Press is an imprint of Taylor & Francis Group, LLC

ISBN: 978-0-367-68086-2 (hbk)
ISBN: 978-0-367-68088-6 (pbk)
ISBN: 978-1-003-13415-2 (ebk)

Typeset in Computer Modern font
by KnowledgeWorks Global Ltd.

# Contents

**Chapter 1**     Introduction ................................................................................. 1


## PART I   *Geomagnetic Disturbances*

**Chapter 2**     Solar Activity as a Danger to Ground-based Technological
                  Systems ..................................................................................... 7

    2.1   The Sun and space weather...................................................... 8
        2.1.1   Cycles of solar activity – Explosive events on
               the Sun ...................................................................... 8
        2.1.2   Solar wind – Density, speed and composition ......... 12
        2.1.3   Magnetosphere – Earth in solar wind ...................... 15
    2.2   Physics of geomagnetic disturbances ..................................... 18
    2.3   Geomagnetically induced currents in ground-based
        technological systems ............................................................. 23
        2.3.1   Power grids ............................................................. 23
        2.3.2   Railway automatics ................................................. 28
        2.3.3   Pipelines ................................................................. 31
    2.4   Ground effects of geomagnetic disturbance modeling ......... 36
        2.4.1   Geomagnetic data................................................... 37
        2.4.2   Geoelectric field modeling..................................... 42
        2.4.3   Geomagnetically induced currents modeling in
               power grids – Discrete system on the continuous
               half-space ................................................................ 49
    2.5   Conclusion .............................................................................. 55
    References.......................................................................................... 55

**Chapter 3**     Geomagnetically Induced Currents Observation........................... 65

    3.1   Methods and equipment for the geomagnetically
        induced current observation.................................................... 65
    3.2   Worldwide observation of geomagnetically induced
        currents .................................................................................. 66

3.3   Geomagnetically induced currents observation in
      north-west of Russia ............................................................. 69
3.4   Conclusion ............................................................................. 74
References .......................................................................................... 74

## PART II   Inside the Power System

**Chapter 4**      Reaction of Power Systems to Geomagnetic Disturbances ........... 81

4.1   Geomagnetically induced current impact on power
      system equipment ................................................................. 82
      4.1.1   Power transformers and autotransformers .............. 85
      4.1.2   Synchronous machines ............................................ 97
      4.1.3   Measurement transformers – Relay protection
              system ..................................................................... 102
      4.1.4   Geomagnetically induced current impact on
              other power system equipment .............................. 106
4.2   Modern approaches of geomagnetically induced
      currents integration into the power grid state calculation ... 109
4.3   Conclusion ........................................................................... 115
References ........................................................................................ 116

**Chapter 5**      Mitigation of Negative Geomagnetic Disturbance Impacts on
                   Power Systems ............................................................................... 121

5.1   Definition and principles of power system resiliency ......... 121
5.2   Forecasting of geomagnetic activity .................................... 128
5.3   Technical actions ................................................................. 136
5.4   Operational procedures ........................................................ 146
5.5   Legislative procedures ......................................................... 155
5.6   Conclusion ........................................................................... 161
References ........................................................................................ 163

## PART III   Developing a View of the Geomagnetic
##            Disturbances Risk

**Chapter 6**      Geomagnetic Disturbance as a Perfect Storm ............................ 175

6.1   Natural hazard impact on power system ............................. 176
6.2   Geomagnetic disturbance as a specific natural hazard ....... 180

6.3   Critical factors ................................................................. 183
        6.3.1    Geomagnetic disturbance parameters ................... 185
        6.3.2    Power system parameters...................................... 188
                    6.3.2.1    Power system architecture..................... 188
                    6.3.2.2    Power system operation state ............... 191
                    6.3.2.3    Power system grounding schemes......... 196
        6.3.3    Power system equipment parameters.................... 197
        6.3.4    Awareness ............................................................ 198
6.4   High risk zones ................................................................ 201
6.5   Economic loss estimation ................................................ 205
6.6   Insurance perspective....................................................... 213
6.7   Conclusion ...................................................................... 218
References....................................................................................... 220

**Chapter 7**   Outlook ................................................................... 233


Glossary ................................................................................................. 235


Index ...................................................................................................... 239

# Foreword

I am delighted to write the foreword for a new book, "Geomagnetic Disturbances Impacts on Power Systems: Risk Analysis & Mitigation Strategies" which deals comprehensively with the threat of solar storms on the world's power systems. I have contributed to this field since the early 1990s and worked on the evaluation of the geomagnetic disturbances (GMDs) measured at the Earth's surface including how they create electric fields in the Earth and couple to large power grids producing quasi-dc currents known as geomagnetically induced currents (GICs). My work was focused on the validation of modeling GICs from particular geomagnetic storms, and during my research I had the pleasure of reading and studying the work of Olga Sokolova and her colleagues who have developed excellent techniques to allow the power industry to evaluate and reduce the risks of GMDs to their networks.

The book begins with a discussion of solar activity that is a cause of risk for all ground-based technological systems. The factors that influence the creation of GICs in power grids, railway systems and pipelines are described. The process that applies for the conversion of the induced magnetic fields (GMDs) to electric fields and currents flowing in long conductors is summarized. In addition, methods are described concerning how to measure of GICs in power networks. While there are now widely available databases of GMD measurements from magnetometers, this book provides a summary of GIC measurements observed in Northwest Russia, some of which are not yet found in available databases.

Once one can establish the level of GICs in particular power networks, the GIC reaction of transformers, synchronous machines, measurement transformers and relay protection methods are described in this book. Given the knowledge of potential equipment effects, one can evaluate the reaction of the operational state of a particular power grid, including the possibility of voltage collapse.

A chapter of this book also deals with the mitigation of impacts that may occur on a power grid, including the evaluation of the resiliency of a grid, the use of forecasting information from space weather observatories, the use of operational procedures, technical protection, and even incentives that might be provided by government organizations to encourage power companies to deal with this low-probability, but high-impact hazard.

The final chapter provides some context to GMDs with respect to other natural hazards, and further describes methods to evaluate a particular grid's risk factors in a straightforward fashion. This is extremely useful to power grid operators, as they are not experts in the field of space weather, but they must be able to deal with its impacts. This is the critical message of this extremely valuable book.

Goleta, California, USA, 2020

*William A. Radasky, Ph.D., P.E.*
*IEEE Life Fellow, IEC Lord Kelvin Awardee*
*Managing Engineer, Metatech Corporation*

# Preface

Major solar storms can occur once every 150 years or so. For us on Earth, space weather can range from the beauty of an aurora borealis to the potentially damaging impact of solar storms on interconnected infrastructure. When a solar storm happens, the Sun shoots out very strong currents of magnetic energy into space. Its manifestation on Earth is called geomagnetic disturbance (GMD). GMDs overall affect operation of multiple systems. The focus of this book is the GMD impact on power systems and interconnected infrastructures. Multiple power system equipment damage from strong GMD can cause a large-scale and prolonged blackout. They can also disrupt many of the critical services on which modern society relies, so placing risk on financial markets, properties, life, and health, etc. The ongoing development of ultra-high voltage grids, on the one hand, and the power system equipment aging, on the other hand, result in the expansion of high-risk zones. Aimed at risk engineers, policy-makers and technical experts, this book seeks to provide an insight into the GMD as a natural hazard and to perform the risk assessment of its potential impacts on the power systems as critical infrastructures. The risk mitigation of GMD impacts on ground-based infrastructure is of crucial value to policy-makers, especially since modern economies are highly depended on the reliable security supply. Contrary to the traditional view on the GMD as a high-impact low-frequency event, the GMD is studied as a "perfect storm". This assumption allows one to enlarge the risk mitigation landscape. The developed guidelines can be directly used by the electric utilities who even may not have a comprehensive understanding of the GMD physics. In the first part of the book, the solar activity as a cause of risk in ground-based technological systems and its appearance on the Earth as Geomagnetically Induced Currents (GICs) are discussed. The second part of the book starts by giving an insight into the reaction of power systems to GMDs followed by discussions on the mitigation strategies aiming at reducing and controlling the risks. The last part deals with the GMD as a perfect storm, the power systems critical factors analysis, the high-risk zones identification and an estimation of economic loss, which is a valuable input for the (re)insurance sector. Thereby, this book provides a full risk assessment tool for assessing power systems confronted with space weather risks.

Zürich, Switzerland, August 2020                                         *Olga Sokolova, Dr. sc.*

# Acknowledgments

The authors of this book owe their gratitude to a very large number of people to whom they would like to express their sincere appreciation. For those not listed here, any errors or omissions are completely our own. The authors wish to thank Dr. William A. Radasky, Metatech Corporation, who shared his deep-rooted concerns on the importance of geomagnetic disturbances (GMDs) impact on power grids analysis, and motivated us to carry out this work. This book relies on the past researches on the GMD manifestation modeling conducted by Dr. Ari Viljanen, Finnish Meteorological Institute, Dr. Antti Pulkkinen, National Aeronautics and Space Administration, Dr. David Boteler and Dr. Risto Pirjola from Natural Resources Canada. We would like to thank Prof. Dr. Volker Bothmer, University of Göttingen, and Dr. Peter Burgherr, Paul Scherrer Institut (PSI), respectively for their consultancies on the Sun-Earth physics and comprehensive assessments of power systems.

We would like to acknowledge all the steps taken by Prof. Dr. Charles Trevor Gaunt, University of Cape Town, Prof. Dr. Mike Hapgood, Rutherford Appleton Laboratory, Dr. Alan W. Thomson, British Geological Survey, Prof. Dr. Edward Oughton, University of Oxford, and Prof. Dr. Daniel Ralph, University of Cambridge, for their activities in raising social awareness of the phenomenon and stressing regional features of its assessment.

Most especially our gratitude goes to Dr. Yaroslav Sakharov, Polar Geophysical Institute, and Dr. Vasilii Selivanov, Kola Science Centre of the Russian Academy of Sciences, for sharing invaluable information on the geomagnetically induced current (GIC) observation in the North-West of Russia, without which Chapter 3 could not exist. A special acknowledgment goes to Prof. Dr. Victor Popov, Peter the Great St.Petersburg Polytechnic University, for his consultancy on the synchronous machines modeling and overall tentative support in the course of this book project.

This work would not be complete without expert guidelines of Swiss Reinsurance Company Ltd (Swiss Re), Zürich-Switzerland.

# Authors

**Olga Sokolova** is a risk analyst with engineering and business administration background. She received her doctoral degree in 2017 from Peter the Great St. Petersburg Polytechnic University (SPbPU). Her research interests lie in the field of critical infrastructure risk assessment to natural and technical hazards. In cooperation with Swiss Reinsurance Company Ltd. (Swiss Re), Swiss Federal Institute of Technology in Lausanne (EPFL), and Paul Scherrer Institute (PSI), Olga Sokolova has been responsible for development and analysis of structural risk-management tools toward a sustainable future which can be used by the general public. This includes a lead-authored brochure published by Swiss Re in 2014 addressing the space weather challenges and opportunities to industry. Dr. Sokolova has a record in raising social awareness of space-borne risks and opportunities brought to society by modern infrastructure developments. She regularly shares her thoughts on natural hazards risks at the corresponding events. In cooperation with the Technology Assessment group at Paul Scherrer Institut (PSI), she contributes in developing ad-hoc risk models validations, challenging the appropriateness of assumptions and modeling processes, and risk governance. In her doctoral research, Dr. Sokolova developed an algorithm for preliminary power grid vulnerability assessment against geomagnetic disturbances.

**Nikolay V. Korovkin** was born in St. Petersburg, Russia in 1954. Professor Korovkin is currently the head of Theoretic Electrical Engineering Department at Peter the Great Saint Petersburg Polytechnic University (SPbPU). He obtained M.Sc., Ph.D. and Doctorate degrees in electrical engineering respectively in 1977, 1984, and 1995. He is an academician of the Academy of Electrotechnical Sciences Russian Federation (1996). Prof N. Korovkin worked in the Swiss Federal Institute of Technology in Lausanne (EPFL), Switzerland (1997), University of Electro-Communications, Japan (1999–2000), and Otto-von Guericke University, Germany (2002–2004). Nikolay V. Korovkin has about 250 published and peer-reviewed works in leading international and Russian journals including the principal book *Theoretical Basics of Electrical Engineering* for the corresponding course in Russian universities. The book consists of three volumes and was published in 2007 in Russia and further translated in China and Vietnam. Together with Prof. M. Hayakawa, he prepared monographs on *Inverse Problems in Electric Circuits and Electromagnetics* (Springer 2005) and *Large AC Machines* (Springer 2017). He studies the inverse problems in electric circuits and electromagnetics, transients in transmission line systems, nonlinear systems, power control systems described by stiff equations, the problems of the electromagnetic prediction of earthquakes, and identification of the behavior of the biological objects under the influence of the electromagnetic fields. His research activities of the last years lie in the field of multiobjective optimization methods analysis and study of the problems associated with renewable energy sources implementation.

**Masashi Hayakawa** was born in Nagoya, Japan on February 26, 1944. He got his MS degree and Doctor of Engineering degree, all from Department of Electrical Engineering of Nagoya University in 1968 and 1974 respectively. He joined the Research Institute of Atmospherics, Nagoya University as a research associate in 1970 and became Assistant Professor in 1968 and Associate Professor in 1969. Dr. Hayakawa worked for Sheffield University (UK) as a visiting lecturer in 1975–1976 and for Laboratoire de Physique et Chimie de l'Environnement (LPCE) (France) as a visiting professor in 1980–1982. Since 1991 he has been a professor of University of Electro-Communications (UEC) and retired from the university in 2009. Then he is an Emeritus Professor of UEC, and he has, in 2011, established a venture company of Hayakawa Institute of Seismo Electromagnetics, Co. Ltd. (Hi-SEM) acting as CEO. He was the Vice-chair and Chair of Commission E of URSI, and the presidents of Atmospheric Electricity of Japan and of Earthquake Prediction Society of Japan. Also he was a co-editor of URSI journal, Radio Science, and is now Editor-in-Chief of *Open Journal of Earthquake Research*. He has authored and co-authored over 800 papers, and have published and edited more than 40 monographs. His interests are simply radio noises just around the Earth, but very diverse; plasma waves in the ionosphere/magnetosphere (their generation and propagation), atmospheric electricity (sferics, Schumann resonances, lightning physics, etc.), seismo-electromagnetics (electromagnetic phenomena associated with earthquakes for earthquake prediction), EMC, signal processing (direction finding, inverse problems).

# Acronyms

| | |
|---|---|
| **AAL** | average annual loss |
| **AC** | alternative current |
| **ACE** | Advanced Composition Explorer |
| **ADC** | analog-to-digital converter |
| **AE** | auroral electrojet (geomagnetic index) |
| **AEP** | aggregate exceedance probability |
| **AGA** | American Gas Association |
| **AU** | astronomical unit |
| **CIM** | complex image method |
| **CIP** | critical infrastructure protection |
| **CIR** | co-rotating interaction region |
| **CME** | coronal mass ejection |
| **COSPAR** | ICSU's Committee on Space Research |
| **DC** | direct current |
| **DGA** | dissolve gas analysis |
| **DKIST** | Daniel K. Inouye Solar Telescope |
| **DMM** | differential magnetometer method |
| **DR** | damage ratio |
| **Dst** | distributed storm time (geomagnetic index) |
| **EASCO** | Earth-Affecting Solar Causes Observatory |
| **ENTSO-E** | European Network of Transmission System Operators for Electricity |
| **EPRI** | Electric Power Research Institute, US |
| **ESA** | European Space Agency |
| **EST** | Eastern Standard Time |
| **ETHZ** | Swiss Federal Institute of Technology in Zurich |
| **EU** | European Union |
| **EURISGIC** | European Risk from Geomagnetically Induced Currents |
| **EUROHOM** | European Resistivity Ohm model |
| **FEMA** | Federal Emergency Management Agency, US |
| **FERC** | Federal Energy Regulatory Commission, US |
| **FMI** | Finnish Meteorological Institute |
| **GIC** | geomagnetically induced current |
| **GIS** | geographic information system |
| **GM** | Geiger-Mueller counter |
| **GMD** | geomagnetic disturbance |
| **GNSS** | global navigation satellite system |
| **GOES** | Geostational Operational Environmental Satellite |
| **GSM** | Geocentric Solar Magnetospheric coordinate system |
| **HF** | high frequency (3–30 MHz) |

**HV**    high voltage
**HVDC**  high voltage direct current
**IMF**   interplanetary magnetic field
**ICME**  interplanetary coronal mass ejection
**ILWS**  International Living With a Star program
**INTERMAGNET**  International Real-Time Magnetic Observatory Network
**IOM**   input-output model
**IZMIRAN**  Pushkov Institute of Terrestrial Magnetism, Ionosphere and Radio Wave Propagation, Russian Academy of Sciences
**JAXA**  Japan Aerospace eXploration Agency
**Kp**    A planetary index of geomagnetic activity that ranges from Kp = 0 to Kp = 9 where Kp = 9 represents the most severe storm
**LASCO** Large Angle and Spectrometric Coronagraph
**LEO**   low Earth orbit
**MLT**   magnetic local time
**MOSWOC**  Met Office Space Weather Operations Centre, UK
**NASA**  National Aeronautics and Space Administration
**NERC**  North American Electric Reliability Corporation
**NOAA**  National Oceanic and Atmospheric Administration
**NRA**   National Risk Assessment
**O2R**   operations to research
**PMU**   phasor measurement unit
**R2O**   research to operations
**s.f.u.**  solar flux unit
**SI**    sudden impulse
**SO**    storm onset
**SOHO**  Solar abd Heliospheric Observatory
**SPE**   solar proton event
**SSA**   space situational awareness
**SSC**   sudden storm commencement
**STEP**  Spare Transformer Equipment Program
**STEREO**  Solar-Terrestrial Relations
**SWORM**  Space Weather Operations, Research and Mitigation Task Force
**SWPC**  Space Weather Prediction Centre
**TEPCO** Tokyo Electric Power Company
**THD**   total harmonic distortion
**TSO**   transmission system operator
**UHV**   ultra-high voltage
**UK**    United Kingdom
**ULF**   ultra-low frequency
**UN**    United Nations
**UNISDR**  United Nations Office for Disaster Risk Reduction
**UPS**   Unified Power System of Russia
**UTC**   Coordinated Universal Time

| **UZH** | University of Zurich, Switzerland |
| **WEF** | World Economic Forum |
| **WMO** | World Meteorological Organization |

# List of Figures

1.1   Risk management paradigm ...........................................................................3

2.1   Schematic drawing of the solar atmosphere structure ...................................9
2.2   Averaged sunspot numbers distribution.......................................................11
2.3   Cumulative distribution of coronal mass ejections speeds for the period
      1996–2015 ...................................................................................................13
2.4   Earth's magnetosphere.................................................................................16
2.5   Geomagnetic field components.....................................................................18
2.6   Magnetic reconnection powering geomagnetic storms and substorms .........19
2.7   Schematic representation of GMD types I and II .........................................22
2.8   Interplanetary, magnetospheric, and ionospheric overview of October
      29–30, 2003 geomagnetic storm period........................................................25
2.9   Sequence of events in North America on October 30, 2003 .........................27
2.10  GIC measured at specific sites of four power systems, located in different
      regions for four disturbed days in November 2004 ......................................29
2.11  Change of signal due to the presence of train wheel ...................................30
2.12  Example of railway automatic failure during a substorm at 18:15 UT
      January 19, 2003 ..........................................................................................30
2.13  Failures occurrence in railway automatics given for three zones at
      different levels of Kp ...................................................................................31
2.14  Pipeline voltages at two sites, time derivative of the horizontal
      component of geomagnetic field and $A_p$ and Dst indices for November
      7–12, 2004 ...................................................................................................34
2.15  Algorithm of ground effects of space weather simulation.............................36
2.16  Horizontal magnetic field inhomogeneity during a
      geomagnetic disturbance ..............................................................................37
2.17  Geomagnetic information nodes....................................................................39
2.18  The ionospheric auroral electrojet current model ........................................43
2.19  Complex image method.................................................................................45
2.20  Example of the Earth structure ....................................................................46
2.21  Geomagnetic induction in power systems ....................................................49
2.22  Five substation power network scheme ........................................................51
2.23  The scheme of **wye$_0$**-Δ connected power transformer and its
      vector diagram .............................................................................................54

3.1   Finnish high voltage power grid models........................................................67
3.2   The GIC measurement network in the North-West Russia.............................70
3.3   The GIC measurement equipment in the North-West Russia.........................71
3.4   Parameters of geomagnetic disturbance on June 29, 2013 ...........................72

3.5  H-component of magnetic field and geomagnetically induced current
     measured on September 7, 2017 event ........................................................... 73

4.1  Principles of GIC distribution in power grid ................................................. 83
4.2  Principles of GIC distribution in power grid in case of the single-phase
     transformer group installation...................................................................... 83
4.3  Pathways to blackout ..................................................................................... 84
4.4  Power transformer construction schemes: core-form transformer and
     shell-form transformer ................................................................................... 87
4.5  Magnetic flux curve shape ............................................................................ 89
4.6  Generalized power transformer magnetic circuit........................................... 90
4.7  Flux distribution due to a high value geomagnetically induced current in
     power transformer as a function of power transformer type........................... 92
4.8  Magnetic flux curve shape ............................................................................ 93
4.9  Low energy insulation degradation triangle for power transformer during
     2003/4 geomagnetic events............................................................................ 95
4.10 Current transformer equivalent circuit......................................................... 102
4.11 Ni-Fe and Si-Fe alloys saturation curves .................................................... 103
4.12 Overcurrent, distance and differential current relay logic schemes............. 105
4.13 Percentage restraint characteristic .............................................................. 106
4.14 Compensation device scheme...................................................................... 114
4.15 Algorithm for GIC impact assessment on power grid state ........................ 115

5.1  The combination of principles for boosting power system resiliency .......... 127
5.2  Complex of measures for preventing negative geomagnetic disturbance
     impacts on power systems ........................................................................... 128
5.3  Representation of SWPC Alerts/Warning for August 2011 coronal mass
     ejections ...................................................................................................... 131
5.4  A pathway of how numerical simulations can be coupled together from
     the Solar surface to the Earth ...................................................................... 133
5.5  Top priority needs to advance understanding of space weather to better
     meet user needs........................................................................................... 135
5.6  Schematic illustration of some concepts in ensemble forecasting, plotted
     in terms of an idealized two-dimensional phase space............................... 136
5.7  Geomagnetically induced currents flow paths ............................................ 137
5.8  Logic scheme for DC-blocking device ........................................................ 138
5.9  Scheme for DC-blocking capacitive device................................................. 140
5.10 Visualization of the April 2000 GMD as observed by National Grid
     Company operator in the control room......................................................... 144
5.11 Equivalent Hydro-Québec power grid ......................................................... 145
5.12 Calculated additional reactive power losses, MVAR, in National grid,
     UK, for various storm scenarios .................................................................. 146
5.13 Calculation of the admissible power flow over the interface....................... 149
5.14 Mitigation square........................................................................................ 155

6.1  Natural hazard types ................................................................................. 178
6.2  Number of relevant natural loss events worldwide 1980–2018 .................... 179
6.3  Sunspot cycles over the last century ........................................................... 181
6.4  Geomagnetically induced current fluctuation in power transformer neutral on substation 330 kV Vykhodnoy during the strong GMD Kp=7 on June 1, 2013 ................................................................................... 182
6.5  Geomagnetic disturbance socio-economic impact pathways ....................... 183
6.6  Group of factors that determine power grid robustness to geomagnetic disturbances ............................................................................................... 184
6.7  Global distribution of the peak geoelectric field determined for the Carrington-event type simulation ............................................................... 187
6.8  Network graph of Central Republic Yakutia power system ........................ 192
6.9  Network graph of Central Republic Yakutia power grid corresponding to summer minimum load normal state .......................................................... 194
6.10 Network graph of Central Republic Yakutia power grid corresponding to summer minimum load state in case of substation 220 kV Suntar loss ........ 194
6.11 Criteria for power equipment quality ......................................................... 198
6.12 Geoelectric amplitudes (V/km) that can be expected to be exceeded once-per-century at EarthScope and USGS magnetotelluric survey sites for north-south geomagnetic variation with a period of 240 s and realized over a duration of 600 s ............................................................................. 202
6.13 Algorithm to find power grid's bottlenecks to negative geomagnetic disturbance impact ..................................................................................... 203
6.14 Graphical visualization of Siberian unified power grid vulnerability to severe geomagnetic disturbance ................................................................. 206
6.15 Primary and secondary critical infrastructure disruptions caused by severe geomagnetic disturbance ................................................................. 208
6.16 Catastrophe model structure ...................................................................... 215
6.17 Hierarchy of the groups of critical factors that define power grid vulnerability to GMDs ................................................................................ 219
6.18 Equivalent 400 kV power system of Scandinavia ....................................... 231
6.19 Scandinavian power system according to geomagnetic latitudes distribution ................................................................................................. 232

# List of Tables

2.1 Basic solar wind characteristics near Earth's orbit .......................................... 14
2.2 Sequence of events: Hydro-Québec blackout March 13, 1989 ...................... 24
2.3 Failures reported in the Swedish high-voltage power transmission system during October 29–31, 2003 storm ................................................. 26
2.4 List of severe magnetic storms when failures have occurred in the automatic system of Northern railway ........................................................ 32
2.5 Largest GIC days in the period 1998–2005 affected gas pipeline at Mäntsälä ............................................................................................. 35
2.6 List of events for magnetic model validation ............................................. 40

3.1 Geographical coordinates of the substations with installed GIC registration equipment ............................................................................... 71
3.2 Maximum geomagnetically induced currents values registered at 330 kV Vikhodnoy substation ................................................................................ 73

4.1 Geomagnetic storm severity according to the Kp and Dst indices ................ 82
4.2 Power transformer repair time as a function of its type and the damage severity ....................................................................................................... 86
4.3 Power transformer's sensitivity to GIC as a function of construction scheme ....................................................................................................... 91
4.4 High harmonic distribution in case of GIC appearance ............................... 94
4.5 List of equations for effective GIC determination ....................................... 96
4.6 The comparison of turbogenerator characteristics as a percentage of the same characteristics for 60 MW model ..................................................... 97
4.7 The value of the main and additional ohmic losses in the stator windings (losses in the Cu) ..................................................................................... 102
4.8 Comparison of power system equipment susceptibility to geomagnetically induced currents .................................................................................... 108

5.1 Summary of resilience practices to geomagnetic disturbances ................... 124
5.2 Characteristics of typical power grid outages and extreme events ............. 125
5.3 Examples of recognized space weather prediction services ....................... 134
5.4 Stability coefficients ............................................................................... 148
5.5 Admissible voltage deviations in PJM power grid ................................... 149
5.6 List of normative contingencies ............................................................. 152
5.7 Characteristics of remedial actions ......................................................... 153
5.8 Example on how to prioritize critical customers and integrate this information in emergency planning ........................................................... 161
5.9 Risk management framework .................................................................. 162

6.1   Overview of natural hazard effects in power systems ................................ 176
6.2   Major outages worldwide caused by natural hazards based on the
      number of affected customers ................................................................ 177
6.3   The characteristics of power system nodes ............................................. 193
6.4   The characteristics of graph nodes ........................................................ 195
6.5   Grid's efficiency as a function of its operation mode ............................. 196
6.6   Overview of blackout in Moscow on May 25, 2005 and St. Petersburg
      on August 20, 2009 .............................................................................. 209
6.7   Catastrophe model functions ................................................................ 214
6.8   Vulnerability functions development approaches .................................... 216

# 1 Introduction

O VER centuries, mankind was fascinated by natural forces, and evolved practices, firstly for worshiping them, then mitigating their negative effects, and
later, most recently, for studying their physics. Efficient or almost efficient mitigation techniques were developed for common terrestrial hazards. Due to high demand
and necessity, these assessment techniques were put into practice. However, space-
borne hazards were treated as a scientific curiosity, since the danger posed by them
was not evident until a certain point. They were looked as invisible neighbors, who
pay occasional or accidental visits, and can just have a cup of tea or rather ask for a
seven-course dinner! The ever-changing list of emerging risks, and reshaping landscape of system interconnections and vulnerabilities drive the need for precise and
advanced estimation of the geomagnetic disturbance (GMD) impact.

The credit of establishing a new branch of science, called geomagnetism, goes to
William Gilbert after publishing De Magnete in 1600 [1]. The main studied question was the geomagnetism impact on the marine navigation. The following steps
supported the research in that era: the creation of the maps of the Earth's magnetic
field declination firstly by Edmund Halley in the beginning of the XVIIIth century,
the discovery of the daily and diurnal magnetic field declination by George Graham
in 1722, and the correlation between large magnetic declination perturbations and
aurora studied by Andreas Celsius in 1741. Nevertheless, the term "Geomagnetic
storm", or "Magnetische Ungewitter" as it was originally named, was introduced by
Alexander von Humboldt when he concluded that the magnetic disturbances on the
Earth's surface and aurora borealis are associated with the same phenomenon. Later,
the connection between sunspot cycle and geomagnetic activity was demonstrated
by Edward Sabine in the middle of the XIXth century. Almost one century later, the
relation between large solar flares and geomagnetic storms was proved as a sufficient
amount of statistical data was gathered.

It is a sobering fact that nowadays natural hazards account for the majority of
economical losses in the world. The ranking of the "major natural hazards" risk done
by the World Economic Forum is constantly emerging over the years. The GMD
is recognized as a new natural hazard of the modern technological age. The societal experience in the XXth century showed the importance of an in-depth study
of the physics of the Sun-Earth system for the economic security and the nations'
well-beings. Several great steps were taken in the last decades towards enriching our
knowledge, which will certainly continue at a faster pace in the upcoming decades.
At the moment of finalizing this book, two remarkable events happened that will
bring us closer to a better GMD assessment. The Daniel K. Inouye Solar Telescope

(DKIST), a brand new facility positioned atop Haleakalā, a 3,000 m-high volcano on the Hawaiian island of Maui, has released pictures that show the Sun surface's features as small as 30 km across in January 2020. The DKIST is a superb complement to the Solar Orbiter (SolO) space observatory which is being launched in the early February 2020 from Cape Canaveral in Florida. The Solar Orbiter is a new mission dedicated to solar and heliospheric physics and will see features as small as 70 km across. Overall, the scientific activities are supported by the improved legislation in national and international scales. There are countries who have recognized the risks associated with severe GMDs and have launched initiatives to prevent, prepare for and respond to this threat. They specified the tasks that will lead to improvements in policies, practices and procedures for decreasing the vulnerability of their critical infrastructures. However, GMD impacts are relatively unfamiliar to the general public.

The matter of GMD became an important issue for the policy-makers, especially since modern economies are highly dependent on the reliable electricity supply. The primary avenue of catastrophic damage caused by the GMD is through the power system infrastructure. Modern infrastructure systems are highly interconnected. A power system malfunction can, in turn, impair the operation of other critical infrastructures. The impact can be both direct or indirect, immediate or delayed. The growth of the population and further economic developments shift the energy consumption towards a higher share of electric power. New technologies and processes development and implementation result in higher cost of electricity undersupply. Moreover, the society expects an increased reliability of power supply and a reduced restoration time.

While procedures for the operation of critical infrastructures assess fairly well the threats from the more common natural hazards, recent efforts to introduce into the picture the GMD as another natural hazard has raised new questions. The goal of this book is to describe the GMD as a natural hazard and to perform its risk assessment. Overall, risk management paradigm consists of five interconnected phases named as analysis, evaluation, research/control, communication, and monitoring. The risks and uncertainties determination is done in the first phase, which are later identified at the evaluation phase. The phase two is also devoted to assessing risks and uncertainty reduction options. The third phase consists of two parallel actions such as research, which targets uncertainties reduction and control, i.e. reducing and controlling the risks. The risks and mitigation management are delivered to the stakeholders within the fourth phase. The ongoing confirmation/revision of assumptions about the risks is made in the last phase. The relevant information for the entire risk assessment steps is presented in this book. A graphical representation of the paradigm is given in Fig. 1.1.

This book has a strong interdisciplinary emphasis, focusing on issues related to risk determination, risk evaluation, and risk management. Each chapter is meant to be understandable relatively independent of the others. Subsequent to this introduction, Chapter 2 starts with an overview of the Sun-Earth physics, followed by the information on the modeling advances. The extended list of historical evidences of

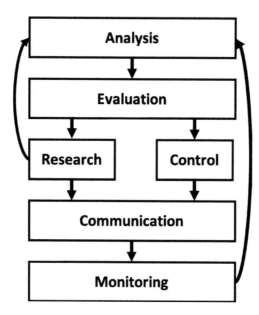

Figure 1.1: Risk management paradigm

negative GMD impacts on terrestrial technological infrastructures such as power systems, railways, and pipelines is also presented here. The observation of GMDs manifestation on the Earth's surface, so-called geomagnetically induced currents (GICs), is a key step in correlating the processes in the heliosphere and the responses to them in the technological systems. Chapter 3 is dedicated to the GIC observation activities around the globe. This chapter furthermore explores the GIC observation in Russia which is not well known compared to other countries. Chapter 4 gives information on power system equipment operation principles and its vulnerabilities to GMD effects as a unit and as a part of the system. Since the GMD risk assessment is a multidisciplinary problem, mitigation activities also should be implemented on several domains. Four domains are distinguished: GMD forecasting, power system hardening, power grid operation algorithm adjustment, and legislation. The maturity of national mitigation strategies against GMD strongly depends on the geographical regions. Countries who have already experienced the negative effects of GMD demonstrate a better performance in case of a severe event. State-of-the-art in each domain together with the list of recommendations regarding the future steps are outlined in Chapter 5. The geography of power systems who are regarded as vulnerable to GMDs is changing. It is not any more the pure problem of high-latitude countries, which were traditionally considered to be susceptible to solar activity manifestation. Chapter 6 shows that the power grid robustness to GMDs is multi-criterial. Critical factors that determine the power grid robustness to GMDs have been the focus of a variety of studies. Nevertheless, their ranking with respect to their impact was not performed up to now. An in-depth analysis of critical factors in particular and as a whole is carried out followed by the introduction of an algorithm for the estimation of the power grid's vulnerability to GMDs. Adequate economic loss estimation is a silver bullet in optimal mitigation principles planning. Chapter 6 is supported by the information on economic loss evaluation associated with an event caused by a severe GMD, and preliminary catastrophe model used by insurance industry.

The authors hope that this book will provide invaluable insights for scientists, engineers, risk analysts, policy-makers, and general public on the GMD impact on the modern economy assessment. The book can be seen as a continuation of discussions initiated in: "Space Weather: Physics and Effects" by Bothmer V. and Daglis I.A. (2007); "Extreme Events and Natural Hazards: The Complexity Perspective" by Sharma A.S., Bunde A., Dimri V.P., and Baker D.N. (2013); and "Extreme Events in Geospace: Origins, Predictability, and Consequences" by Buzulukova N. (2017).

## REFERENCE

1. Gilbert, W., Short, P. (1600). *Guilielmi Gilberti Colcestrensis, medici londinensis, De magnete: magneticisque corporibus, et de magno magnete tellure; physiologia noua, plurimis & argumentis, & experimentis demonstrata*. Londini: Excudebat Petrus Short, anno. [PDF]. Retrieved from the Library of Congress, https://www.loc.gov/item/04001021/ (Accessed on August 2020)

# Part I

*Geomagnetic Disturbances*

# 2 Solar Activity as a Danger to Ground-based Technological Systems

## CHAPTER CONTENTS

2.1 The Sun and space weather ............................................................................ 8
  2.1.1 Cycles of solar activity – Explosive events on the Sun ...................... 8
  2.1.2 Solar wind – Density, speed and composition ................................... 12
  2.1.3 Magnetosphere – Earth in solar wind ................................................ 15
2.2 Physics of geomagnetic disturbances ......................................................... 18
2.3 Geomagnetically induced currents in ground-based technological systems ... 23
  2.3.1 Power grids ......................................................................................... 23
  2.3.2 Railway automatics ............................................................................ 28
  2.3.3 Pipelines ............................................................................................. 31
2.4 Ground effects of geomagnetic disturbance modeling ............................... 36
  2.4.1 Geomagnetic data .............................................................................. 37
  2.4.2 Geoelectric field modeling ................................................................ 42
  2.4.3 Geomagnetically induced currents modeling in power grids –
        Discrete system on the continuous half-space ................................. 49
2.5 Conclusion ................................................................................................... 55

FOR us on Earth, solar activity manifestations can range from the beauty of an aurora borealis to the potentially damaging impact of GMDs and the significant risk they pose to our modern society, heavily dependent as it is on technology. The growing appreciation of this problem brings the need for the proper modeling of the solar activity manifestations and their effects on modern society's well-being. This chapter presents the knowledge about various facets of solar activity, and how to model its impact on terrestrial technological systems. The chapter starts with the information on Sun-Earth physics and how GMDs appear. It is followed with the distinguished overview of evidence when GMDs disturbed ground-based technological systems operation such as power grids, railways and pipelines. The modeling principles and aspects are then discussed.

## 2.1   THE SUN AND SPACE WEATHER

### 2.1.1   CYCLES OF SOLAR ACTIVITY – EXPLOSIVE EVENTS ON THE SUN

Arthur Eddington wrote in 1926 that at first sight it would seem that the deep interior of the Sun and stars is less accessible to scientific investigation than any other region of the universe. Our telescopes may probe farther and farther into the depths of space; but how can we ever obtain certain knowledge of that which is hidden behind substantial barriers? What appliance can pierce through the outer layers of a star and test the conditions within? [1]. In other words, Arthur Eddington encouraged to use our knowledge of basic physics to determine Sun's properties instead of developing missions to probe the Sun. The same approach is predominant today. According to modern heliophysics, the Sun is a sphere of high temperature ionized gas (90% hydrogen and 10% helium) with a radius of $R_S = 6.96 \times 10^8$ m which is 109 times larger than the Earth's radius and with a mass of $M_S = 1.99 \times 10^{33}$ g which is 330,000 times heavier than Earth [2]. The surface of the Sun is $6.087 \times 10^{18}$ m$^2$. The average distance from the Sun to the Earth is 149.6 million kilometers, which stays for 1 astronomical unit (AU). The AU is used for measuring the distances in solar system. The Sun rotates with a synodic period of 26.24 days, which is the time for a fixed feature on the Sun to rotate to the same apparent position as viewed from Earth. The direction of rotation is the same as that of Earth and also coincides with the direction of the Earth's rotation around the Sun.

The Sun emits energy in the form of electromagnetic radiation. The information about updated solar irradiance spectrum is presented in [3]. The nuclear reactions determine the processes of energy release and chemical elements formation. The main source of all ionizing and penetrating radiations is the solar atmosphere, which consists of three areas as depicted in Fig. 2.1:

- **Photosphere**: the region closest to the Sun's surface; the lower layer of the solar atmosphere with a thickness 100–300 km. The photosphere emits almost all of the energy emitted by the Sun into space. Maximum of photosphere's radiation is in the visible part of the spectrum. This is due to the fact that the effective Sun's temperature, i.e. the temperature corresponding to the temperature of the absolutely black body, is 5,780 K. The ionizing radiation of the Sun is formed in regions located above the photosphere, primarily in a chromosphere.
- **Chromosphere**: the second of the three main layers of the Sun's atmosphere. It is about 10.4 km thick. Its temperature increases with height from photospheric values to about $10^6$ K. Between the chromosphere and the corona, a transition zone is distinguished.
- **Corona**: a very rarefied outer part of the solar atmosphere. Corona is observed in all areas of the electromagnetic radiation spectrum – from X-ray to radio wave. However, its radiation in the visible region is very weak. Corona continuously expands due to the solar pressure gradient in the opposite direction to the solar gravity into the interplanetary vacuum. This steam of particles is called solar wind.

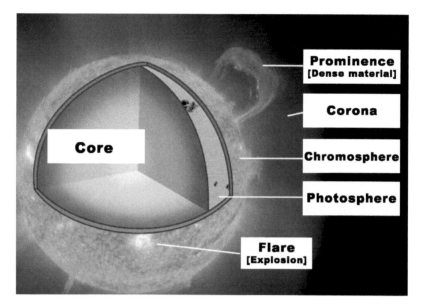

Figure 2.1: Schematic drawing of the solar atmosphere structure (modified from [4]) [Courtesy of NAOJ, Credit: JAXA]

The crucial condition for solar activity existence is the Sun's magnetic field, which is similar to large-scale Earth's dipole field. However, the Sun's magnetic field does not have a well-defined axis and is not symmetrical. Therefore, it cannot be considered that it is created by a dipole inside the Sun. Most likely, it is the result of numerous small surface magnetic fields superposition. The Sun's photosphere magnetic structure varies continuously, hence transient processes (micro- and nano flares, shock waves, erupting prominences, flares and coronal mass ejections (CME)) cause a short-term increase of Sun's emissions. Since intensity variations at UV- and EUV wavelengths may have important effects on the Earth's atmosphere, this subject is in the focus of many research activities [5].

The main feature of the Sun's magnetic field is a cyclical change of its polarity with a period of 22 years which, in turn, is the cause of 11-year solar cycle. Solar cycle is the nearly periodical (11 years) change in the solar activity. The long-term behavior of the 11-year solar cycle variations and that of even higher periodicities (e.g., the Gleissberg cycle of ∼90 years) is highly important in terms of space climate [6]. The most common solar activity measures is the sunspot number. The sunspot is an area with reduced surface temperature which expends less energy. Sunspots can be resembled as a visible darker areas on the Sun's surface. Sunspots are magnetic in nature. Therefore, solar maximum and solar minimum refer to periods of maximum and minimum sunspot counts. Cycle spans from one minimum to the next.

The sunspot's brightness in visible light is 20–30% of the surrounding unperturbed photosphere brightness. The temperature in the center of the spot is 1,600 K lower than the usual one. Sunspots are a kind of gigantic cavities 100 km deep. Their diameter varies greatly from spot to spot and can be up to 50–70 thousand kilometers. In comparison, the Earth's diameter is only 13,000 km. On average, the lifetime of a sunspot group is 10 days. Over a full 11-year cycle, up to 3,000 sunspot groups are observed. All of them have strong magnetic fields with opposite polarity at neighboring spots.

While sunspots are only an indirect proxy for the solar magnetic field central to the solar cycle, they are by far the longest series directly observed solar parameter and thus remain of great scientific interest across a range of research areas [7]. As a quantity to measure the number of sunspots and group of sunspots on the Sun's surface, the Wolf's number $R$ is used. It is also called Zurich sunspot number, since the idea of computing sunspot numbers was proposed by Rudolf Wolf, Professor of Astronomy at the University of Zurich (UZH) and the Swiss Federal Institute of Technology in Zurich (ETHZ), in 1848. It is calculated as presented in Eq. 2.1:

$$R = k(10g + f), \tag{2.1}$$

where $g$ is the number of sunspots groups, $f$ is the number of individual sunspots, $k$ is the coefficient that varies with location and instrumentation (also known as the observatory factor).

The Wolf number is not the true result of observation. It is a semi-quantitative indicator of the "spottedness" of the Sun. Despite this drawback, it turned out that the Wolf number is a good index for describing the solar activity level. Ongoing development of the Sun observation techniques brought new indices for solar activity evaluation. One of them is the solar radio flux at 10.7 cm (2,800 MHz) – F10.7 index. It is measured since 1947, first in Ottawa, Canada, later in the Penticton Radio Observatory in British Columbia, Canada, and it is one of the longest running record of solar activity. It can be reliably measured on a day-to-day basis and is reported in solar flux units (s.f.u.). A good linear correlation between Wolf number and F10.7 index exists [8].

The solar cycle numbering scheme was introduced by R. Wolf and the 1755–1766 cycle is traditionally numbered "1". Currently, we are on the edge of the 24th and 25th solar cycles. The actual duration of the 11-year cycle can vary from 10.0 to 12.1 years according to the data collected over 240 years of observations. The average time interval between the minimum and maximum is 4.3 years and between the maximum and minimum is 6.6 years. Cycles with a shorter growth phase usually have a higher maximum. The cycle height determined by the smoothed monthly averaged sunspot numbers changes more than five times (see Fig. 2.2). The practical interest of solar activity influence on man-made technical systems was prompted in solar cycle 9 (1843–1854) and the investigation of solar activity impact on power grids was emphasized in solar cycle 17 (1931–1943).

The solar cycle change is provoked by the interaction between the Sun's rotation and its magnetic field. The Sun rotates differentially, i.e. the highest rotation speed is observed at the equator and the speed on the poles is about 20% lower. At the beginning of the 11-year solar cycle, the Sun's integral magnetic field may be treated as a magnetic dipole. First, the polarity of the Sun's magnetic field is positive – that

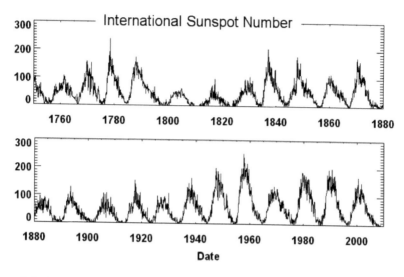

Figure 2.2: Averaged sunspot numbers distribution [9] (Courtesy of Spaceweather.com)

is, the direction of the magnetic field lines is predominantly directed away from the Sun's photosphere – at its northern heliographic pole. Later, the polarity of the Sun's magnetic field is changed to negative, i.e. the direction of the magnetic field lines is predominantly directed towards the photosphere at the Sun's southern heliographic pole. A complete magnetic polarity reversal of the Sun's magnetic field takes about 22 years, commonly referred to as the "Hale Cycle" [10].

Sunspots normally appear in active regions which are characterized by magnetic field anomalies. A relatively small local field strength increase results in an increased energy loss of waves coming from the convection zone. In the corresponding zones of the chromosphere, there is an increase in brightness in the lines of calcium and hydrogen. These zones are called flocculi. Approximately in the same areas of the photosphere, but somewhat deeper, there is also an increase in brightness in visible (white) light. These zones are called torches. Torches are 200–300 K hotter than flocculi.

Local magnetic fields in active regions cause intensive movement of large masses of ionized gas particles in the form of prominences. A comprehensive understanding of solar prominences ultimately requires complex and dynamic models, constrained and validated by observations spanning the solar atmosphere [11]. The stable prominences are mainly observed. They are seen as long dark stripes of irregular shape, fibers, on the disk of the Sun. Huge eruptive prominences are observed as a huge sparkling arcs or loops that arise above the Sun's surface. The example of solar prominence from June 4, 1946 called "granddaddy" is shown in [12].

The development of space-borne observation methods allowed us to discover the coronal holes, i.e. area of the Sun's corona lower-density plasma. The plasma temperature is also lower (1.5–2 times). They are the regions of open, unipolar magnetic field. Therefore, the outflow of the corona material restrained by the magnetic field in other regions can occur in the holes. The coronal holes are constantly changing and reshaping since the corona is not uniform. Coronal holes are associated with the appearance of the so-called high-speed solar wind flows, which play an important role in GMDs formation. This open magnetic field line structure permits the solar wind to escape into space. It is accepted that the appearance of coronal holes on the solar disk leads to a solar wind increase. The most dramatic variations in ionizing radiation in near-Earth space are CMEs which are large expulsions of plasma and magnetic field from the Sun's corona. They can eject billions of tons of coronal material and carry an embedded magnetic field that is stronger than the background solar wind interplanetary magnetic field strength. CMEs travel outward from the Sun at speeds ranging from slower than 250 km/s to as fast as near 3,000 km/s. Two types of closed field regions are known to produce CMEs: sunspot regions (active regions) and quiescent filament regions [13]. The fastest CMEs originate from active regions. Magnetic energy possessed by active regions is higher, and it depends on active region size and the average field strength. The potential energy in active region can be estimated as a measure of the maximum free energy available to power eruptions. The distribution of solar sources of CMEs as a function of GMD's intensity is presented in [14]. Strong CMEs originate within a central meridian distance of about $20°$.

One of the CMEs characteristics is their speed. The distribution of CME speeds ($V$) measured by Large Angle and Spectrometric Coronagraph (LASCO) installed on SOHO is shown in Fig. 2.3. The used abbreviations are following: m2 – metric type II radio bursts due to shocks in the corona without interplanetary extension, MC – CMEs associated with magnetic clouds detected at 1 AU, EJ – CMEs associated with interplanetary CMEs lacking flux rope structure, S – CMEs associated with interplanetary shocks detected at 1 AU, GM – CMEs associated with major GMDs (Dst$\leq -100$ nT), Halo - halo CME population, DH – CMEs associated with decameter-hectometric type radio II bursts, SEP – CMEs associated with large solar energy particles (SEP) events, GLE- CMEs associated with ground level enhancement in SEP events. The fastest CME ($V \sim 3,380$ km/s) in Fig. 2.3 happened on November 10, 2004 at 02:26 UT. It was concluded that the initial high speed was likely to be due to the propelling force (solar source property) rather than the drag force (ambient medium property) [15].

## 2.1.2  SOLAR WIND – DENSITY, SPEED AND COMPOSITION

The solar wind discovery is mainly associated with the name of Eugene Parker. The history of discovery is presented in details in [17]. The observed steam of charged particles using spaceflight instrumentation showed a systematic two-stream pattern and a 27-day recurrence interval. It allowed us to suggest that their solar source regions are rotated with the Sun [18].

Solar wind is a steam of charged particles emitted from the Sun's corona. The flow of solar wind has been studied for over 40 years until today, and an abundant amount of data has been accumulated concerning its average properties and

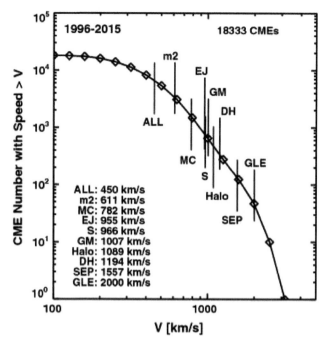

Figure 2.3: Cumulative distribution of coronal mass ejections speeds for the period 1996–2015 [16]

intermittent energetic manifestations, which impact the Earth's magnetosphere [19]. Different regions on the Sun produce solar wind of different speeds and densities. Two types of solar wind are historically accepted: fast and slow, though their most distinctive property is not the solar wind speed, but the elemental and charge state compositions. There have been a number of schemes applied to separate different states of the solar wind and attempted to find a best way to distinguish the slow solar wind from the fast solar wind [20]. For instance, the heavy ion charge state ratios ($C^{+6}/C^{+5}$ and $O^{+7}/O^{+6}$) have been used as a classification scheme [21].

The slow solar wind has an average speed of 400 km/s, a temperature of ca. 105 K and a composition that is a close match to the corona. This average speed corresponds to calm conditions when there are no flares and no coronal holes. However, solar wind slower than 300 km/s is termed very slow solar wind, which is seldom observed at 1 AU [22]. The coronal holes appearance is accompanied by the injection into the interplanetary space of the coronal mass with velocities 1.5–2 times higher. This solar wind is named to be fast. It has a typical velocity of 750 km/s, a temperature of $8 \times 10^5$ K and its composition nearly matches the Sun's photosphere composition. Unlike the slow solar wind, it is not evenly distributed in the ecliptic plane, but concentrated in areas that look like cones expanding in the direction from the Sun, whose projection on the Earth's orbit is 40–60° N. The overview of solar wind characteristics is given in Table 2.1 (this table is adapted from [23]).

**Table 2.1**
**Basic solar wind characteristics near Earth's orbit**

| | Fast solar wind | Slow solar wind |
| --- | --- | --- |
| Flow speed | 450–800 km/s | $<\sim$450 km/s |
| Proton density | $\sim$3 cm$^{-3}$ | $\sim$7–10 cm$^{-3}$ |
| | $\sim$95% H, 5% He, minor ions and same number of electrons | $\sim$94% H, 4% He, minor ions and same number of electrons – great variability |
| Proton temperature<br>B $\sim$5 nT | $\sim$2$\times$10$^5$ K<br>B $\sim$4 nT | $\sim$4$\times$10$^4$ K |
| | Alfvénic fluctuations | Density fluctuations |
| | Origin in coronal holes | Origin above coronal streamers and through small-scale transients |

More powerful fast solar winds are associated with the release of huge masses of matter from the chromosphere and corona. Their speed can exceed 1,000 km/s. The projection of their cones in the Earth's orbit can be 1.5–2 times larger and reach 120°. The GMDs, which are the focus of this book, are associated with these ultra-fast Earth-directed solar winds.

The important solar wind feature is that, in addition to particles (protons and electrons), they also carry away a magnetic field from the Sun. The pattern of solar wind particles distribution in interplanetary space is complex. Magnetic fields are "frozen" into the solar wind plasma. The trajectory is close to the Archimedean spiral as viewed from above of the ecliptic plane. It is also termed as Parker spiral. This is due to the fact that the ejection of coronal matter does not occur from stationary, but from rotating gas balls. A similar pattern can be observed for the jets in the sprinkler. The solar wind plasma moves in a spiral, therefore the lines of force of the interplanetary magnetic field also have the spiral form. The inclination angle of the spirals relative to the radial direction in the Earth's orbit is ca. 45° for a flow speed of 400 km/s. The angle of the magnetic field direction $\phi$ is calculated as in Eq. 2.2:

$$\phi = \left( \frac{\Omega_S R}{V_R} \right), \tag{2.2}$$

where $\Omega_S$ is the Sun's rotation speed and $V_R$ is the solar wind speed at a given distance $R$.

The interplanetary magnetic field has a sector structure – the "Slavgaard-Mansurov effect". The magnetic field differs from sector to sector in direction from the Sun (plus) and to the Sun (minus). The sectors do not necessarily have the same size. Some pass through the Earth within 8 days, others within 4 days. The sectorial structure and polarity of the interplanetary magnetic field components determine the nature of the interaction of the interplanetary field and the Earth's magnetic field with all the ensuing consequences. Constant monitoring of the ionized and penetrating radiations of the Sun are performed with Geostationary missions.

### 2.1.3 MAGNETOSPHERE – EARTH IN SOLAR WIND

As some other planets, Earth has its own magnetic field called geomagnetosphere – the region of space surrounding Earth where the dominant magnetic field is the magnetic field of the Earth, rather than the magnetic field of interplanetary space. In other words, this is the region where the character of charged particles motion is determined by the Earth's magnetic field. Its size and shape are controlled by three components of the solar wind pressure: the pressure of the interplanetary magnetic field, thermal pressure of the plasma that is measured in the frame of the solar wind flow, and the momentum flux of the cold stream of ions flowing radially from the Sun. The solar wind interaction with geomagnetosphere can be studied using the fluid approach due to the fact that ion's inertial length is infinitesimal compared to stand-off distance of the magnetopause. The credit for the most notable corresponding research belongs to [25].

The observations showed that the geomagnetosphere has a teardrop shape which is the direct result of being blasted by solar wind. The head of the drop (sunward side) $L_{mp}$ extends up to 10 Earth radii toward the Sun. It comes from the following conclusion. A solar wind density $n_{sw}$ is 9.6 protons per cm$^3$, its speed $u_{sw}$ is 400 km/s. Applying these numbers to Eq. 2.3 gives the aforementioned result.

$$L_{\mathrm{mp}} = 107.4 \left( n_{sw} u_{sw}^2 \right)^{-1/6}, \tag{2.3}$$

where $n_{sw}$ is protons per cm$^3$ and $u_{sw}$ is the solar wind speed in km/s.

The tail of the drop stretches away in reaching beyond the Moon's orbit. The magnetotail (no.6 at Fig. 2.4) (night side) extends more than 100 Earth radii (600,000 km) from the Earth. The complex structure of magnetosphere is shown in Fig. 2.4.

Depending on the orientation to the Earth's surface, three types of magnetic field lines (no.7 at Fig. 2.4) can be distinguished:

- Magnetic field lines that do not cross the Earth at all, they stay for interplanetary field lines.
- The magnetic field lines that intersects the Earth's surface twice (by leaving from the south polar region, moving across the equatorial region, and crossing the Earth's surface in the north polar region) are named closed.
- Open magnetic field lines are those that link magnetic field lines in the solar wind by leading the Earth's surface. The third type is responsible for a magnetic reconnection mechanism.

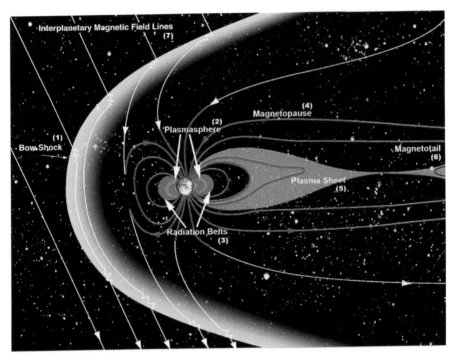

Figure 2.4: Earth's magnetosphere [Credit: NASA/Goddard/Aaron Kaase]

The rate at which magnetic field in the solar wind reconnects with the geomagnetic field depends on the angle between the solar wind and the Earth's magnetosphere. Since the angle is constantly changing, the level of magnetosphere convection also changes. In other words, it determines the rate of transport of reconnected closed magnetic flux from dayside magnetosphere to the tail. The strong interaction occurs when the interplanetary magnetic field is southward, since southward interplanetary magnetic field conditions provide much stronger energy of mass, momentum and energy than northward conditions do. The rate of the process depends on the geometry of reconnection region. In case of a southward interplanetary magnetic field, reconnection region can be up to the entire dayside magnetopause. The region is small in the opposite situation leading to the relatively weak magnetic field strength. The interaction between the solar wind and the Earth's magnetic field, and the influence of the underlying atmosphere and ionosphere, creates various regions of fields, plasmas, and currents inside the magnetosphere such as the plasmasphere (no.2 at Fig. 2.4), the ring current, and radiation belts (no.3 at Fig. 2.4).

These fluctuations make it possible to energize and trap particles in the inner portions of the magnetosphere, creating the radiation belts [26]. The magnetosphere has a boundary along which an electric current known as ring current flows due to an interaction between solar wind ions and magnetosphere. The geomagnetic field deflects the solar wind ions to the west and electrons to the east. The magnetic field of the ring current divides the magnetosphere from the magnetopause – the boundary in space that separates the region dominated by the Earth's magnetic field (the magnetosphere) from the surrounding solar wind [27]. The solar plasma flows in magnetopause (no.4 at Fig. 2.4). The supersonic flow of solar plasma at the moment of reaching the obstacle in the form of magnetosphere forms a bow shock (no.1 at Fig. 2.4). The front of this wave is located in front of the magnetopause at a distance of several tens of thousands of kilometers. This region is called transition region and characterized with high temperature plasma (up to tens of millions of degrees). The solar wind pressure deforms the geomagnetic field. On the day side of the magnetosphere, the force lines are compressed (stronger compression is observed at high latitudes). In contrary, the force lines are swelled on the night side in the way that those which originate from the polar regions are drifted by the solar wind and form a long tail.

It is accepted to divide magnetosphere into three regions:

- **Inner magnetosphere (ionosphere)**: the region of a relatively stable field determined by the inner sources of the Earth. The magnetic field is quasi-dipole in nature. The force lines exit from the South magnetic pole and enter the North magnetic pole. At any point, the geomagnetic field is characterized by a complete vector sum of tension $\overline{T}$. Geomagnetic field components are presented in Fig. 2.5. Its magnitude and direction are defined by three components: X, Y and Z (northern, eastern and vertical) in a rectangular coordinate system or by three elements of geomagnetic field: the horizontal component of magnetic field strength H, magnetic declination D – angle between H and the plane of the geographic meridian) and magnetic inclination I (angle between $\overline{T}$ and horizon plane).

- **Outer magnetosphere**: the region adjacent to the magnetopause. The magnetic field in this region is determined by currents at the boundary of the magnetosphere, and therefore, substantially responds to changes in external conditions. The outer magnetosphere includes magnetotail, boundary layer, and daytime polar cusps (northern and southern). Cusps are funnel-like formations that separate lines of force on the day and night sides of the magnetosphere.

- **Aurora magnetosphere**: the region that divides inner and outer magnetosphere. Its size is determined by the solar wind. The auroral magnetosphere includes the region of closed tail lines at a distance of at least 60 Earth's radii.

At the heights less than 300 km from the Earth's surface, significant local geomagnetic field deviations from the the dipole field are observed. The small-scale inhomogeneities are due to the magnetic ores occurrence in the Earth's core. Large-scale (global) inhomogeneities are associated with the structural features in the deep layers of the planet. Brazilian magnetic anomaly and Eastern Siberian anomaly are the places where the lowest and highest field strengths are observed, respectively. Geomagnetic field strength undergoes periodic and irregular changes of varying

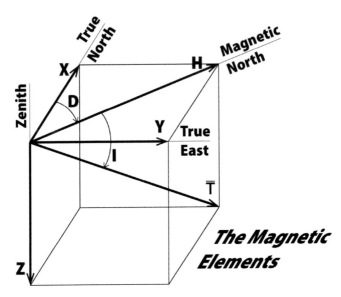

Figure 2.5: Geomagnetic field components (adapted from the NOAA original sketch)

magnitude and duration. The main geomagnetic field caused by the inner sources experiences slow secular changes. The exact information of geomagnetic field parameters can be found using magnetic maps.

Solar wind plasma flow around the magnetosphere with a variable density and velocity of charged particles together with the breakthrough of particles into the magnetosphere leads to the change of electric current system intensity in the Earth's magnetosphere and ionosphere. Current systems, in turn, cause geomagnetic field oscillations in the near-Earth space and on the Earth's surface in a wide range of frequencies ($10^{-5}$–$10^{0}$ Hz) and amplitudes ($10^{-3}$–$10^{-7}$ Oe).

One of the magnetosphere responses types is called substorm, which is caused by both dayside and nightside reconnection. It has both magnetospheric and auroral manifestations. The basic nature of substorm can be understood in terms of the overall Dungey model, modified by the expanding/contracting polar cap model, and the near Earth neutral line model [28]. Substorm is a cycle of energy storage and release consisting of three phases: growth (lasts up to 1 hour), expansion (20 minutes), and recovery (2–3 hours) [29]. When solar wind contains magnetic and plasma structures that impact magnetosphere over hours, such a response is called storm (geomagnetic disturbance).

## 2.2  PHYSICS OF GEOMAGNETIC DISTURBANCES

The Carrington event of 1859 remains as a historical extreme event against which many GMDs are compared. It took place in the time, when extensive observations from magnetometers and global aurora existed. The cause of GMD in solar activity was recognized in 1852 by Sabine as a synchronous variation of sunspot number and geomagnetic activity. Nevertheless, GMDs are acknowledged since the 1600s [29].

GMDs are known to be "solar echo on Earth". It is a disturbance of Earth's magnetosphere caused by an energy exchange from the solar wind into the space environment around the Earth. The intense energy release from the Sun over a relatively short period of time results in generation of shock waves and the release of plasma clouds into the interplanetary space. Whenever an interplanetary shock wave approaches the Earth, the geomagnetosphere is getting compressed and currents in magnetopause are getting increased. Such sudden magnetic field growth covering the entire globe is considered to be the start of a GMD. Each GMD is unique, though some common features can be noted (see Section 6.2).

Solar wind speed is one of the drivers of geomagnetic activity, but mainly the energy transfer into the geomagnetosphere depends on whether the interplanetary magnetic field has a southward-directed (antiparallel) component with respect to the ecliptic plane (magnetospheric field direction). This component, which determines magnetic reconnection between interplanetary magnetic fields and magnetospheric fields, is referred as $B_z$. The idea of Geocentric Solar Magnetospheric (GSM) coordinate systems was introduced by [31]. The process is depicted in Fig. 2.6. The interplanetary magnetic field directed southward is carried by the solar ejecta. As a result of reconnection, magnetic field lines are eroded on the dayside magnetosphere and transported to the nightside magnetotail region, where magnetic field reconnect. The reconnection process results in plasma injection into the nightside magnetosphere. The energetic protons and electrons of the injected plasma move to the west and to the east by forming a ring current around the Earth.

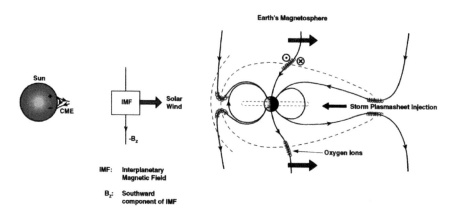

Figure 2.6: Magnetic reconnection powering geomagnetic storms and substorms [32]

GMDs are characterized with a system of indices. The most familiar is the overall geomagnetic activity index Kp (Kennziffer Planetarisch). Similar to Richter scale for earthquakes, it is a quasi-logarithmic parameter, from 0 (quiet) to 9 (extreme), given on a 3-hour basis measured on the basis of observations from 13 geomagnetic stations (11 in the northern and 2 in the southern hemisphere). The imperfection of Kp for describing severe storms is discussed in [33]. The geomagnetic planetary $A_P$ is the cousin of Kp, and it reflects a daily average level for geomagnetic activity. $A_P$ ranges from 0 to ca. 400. The intensity of ring current is measured by Disturbance Solar Storm index Dst, which is derived from a network of near-equatorial geomagnetic observatories. Both Kp and Dst are measured in nano Tesla (nT). Negative Dst values indicate GMD. The empirical equation that correlates the GMD strength Dst and the speed $V$ with which CME magnetic field impinges on the magnetosphere, and the strength of the CME magnetic field component oriented in the direction opposite that of Earth's magnetic field $B_z$ is described in [34].

Both CMEs and co-rotating interaction regions can cause GMDs. More precisely, only space weather manifestations that contain an interplanetary magnetic field component perpendicular to the ecliptic plane can induce the GMD on Earth. These manifestations include: the interplanetary CMEs (ICMEs) including magnetic cloud (MC), non-MC Ejecta, sheath, which is a compression region before the fast leading edge of ICME, and co-rotating interaction regions (CIR) caused by fast solar wind streams catching up slower solar wind streams ahead that had originated in solar longitude westward of the fast streams as viewed from Earth. So-called CME-induced storms are the results of complex phenomena including two geoeffective parts: Sheath and ICME, and the majority of extreme storms are generated by interacting Sheath/ICME structures [35].

Recurrent GMDs are caused by fast solar wind streams with speeds in the range of 500–800 km/s [36]. Recurrent geomagnetic storms are dominant especially in the declining phase of sunspot cycles [37], because large polar coronal holes exhibit persistent low-latitude extensions over several months interval at these times [38]. Geomagnetic activity peaks within the CIR due to the compression and fluctuations of the interplanetary magnetic field (IMF) and Alfvénic waves [39]. The CIR-related activity and subsequent wave activity are the reasons for the typically observed two-step behavior in recurrent storms [5]. The strength and duration of a GMD caused by CIR and related high-speed solar wind stream can be quite variable, depending on the amount of compression of the IMF and the direction of the $B_z$ component as well as its duration in case of a southward direction, and, finally, the spatial size of the following high-speed flow [40]. The storm strength can also depend on solar wind density, but the effect is not significant for extreme storms [41].

GMDs cycle consists of the following phases. GMDs have a main phase during which the horizontal (H) component of the Earth's low-latitude magnetic fields is significantly depressed over a time span of up to a few hours followed by its recovery, which may extend over several days [42]. It is accepted to consider that the main phase starts when the plasma cloud from the Sun reaches the magnetosphere. This phase is characterized by a sequence of explosive processes called substorm associated with the energy release from the magnetotail and its increase into the high latitude ionosphere. The substorm is seen as a sudden brightening of auroral arcs. This process, known as re-polarization, results in the energization of charged particles in the plasma sheet and their injection deeper into the inner magnetosphere.

Several theories exist regarding the substorm expansion. The initial phase is also referred to as a storm sudden commencement (SSC). Nevertheless, not all GMDs have an initial phase as not all sudden magnetic field strength increase is followed by a GMD. The recovery phase of GMD begins when the ring current starts to decay.

Two categories of GMDs are distinguished based on how they start. GMDs without any initial phase are called gradual geomagnetic storms (*Type II*). Other storms are characterized by a sudden increase of the horizontal magnetic field caused by fast ICMEs which compresses the magnetosphere leading to a sudden increase in magnetic field strength (*Type I*). High-speed shocks that arrive at Earth in less than one day are known as fast-transient events [43]. These shocks are considered to be extreme events because they can cause high levels of energetic storm particles at Earth and compress the magnetosphere observed as sudden impulse (SI) or SSC of GMD [44]. Depending on the initial phase differences, the main phase of GMD can also have two behavioral scenarios. The main phase of *Type II* GMDs consists of two-step due to the two-steps ring current growth. The mechanism is following: a new major particle injection occurs before a ring current has decayed to a significant pre-storm level. It corresponds to arrival of the compressed southward IMF in the sheath region downstream of the ICME shocks followed by the southward fields of magnetic cloud [45]. *Type I* GMDs are classified with one-step in main phase, since the ring current is intensified by the magnetotail injection of energetic particles and decays to pre-storm level in one step. In addition to the first two types, *Type III* exists, which can have three or even more steps of the main phase, depending upon the solar and interplanetary conditions, multi-ring current injections [46]. The character of GMD types is presented in Fig. 2.7 (this figure is adapted from [38]). The abbreviation SO stands for storm onset, i.e. onset of the storm main phase.

Generally, ICME-driven storms have higher intensities (intense to superintense storm level) and shorter durations as compared to CIR-driven storms (moderate to intense storm level) intensities and durations [47]. Observations show that the intensity of CIR-driven GMDs does not exceed Kp equal to 8. GMDs characterized by Kp equal to 9 and 9+ are driven by interplanetary shocks associated with ICMEs. The storm is considered to be superintense if $Dst \leq -500$ nT. Only one true superintense storm was registered after the start of space era – Hydro-Québec event in 1989, when Dst reached the value of 589 nT [49]. Space era started in in 1957 with the launch of the first satellite. Halloween event in 2003 had intensity of $Dst \sim -490$ nT. However, the geomagnetically induced currents (GICs) intensity, which is of interest to power grid owners and operators, is controlled by the ionospheric currents rather than the ring current intensity (Dst). However, the GMDs caused by CIRs have long recovery phase up to several weeks [48].

Despite the need to perform permanent GMD forecast, certain dependences on GMD occurrence exist. GMD occurrence has a pronounced semiannual variation. Most of intense GMDs appear at times near equinoctial months (March, September). Clear depression in the number of GMDs is observed during the solstitial months (June, December). The dependence between GMD occurrence and the solar cycle sunspot number is as the following. The GMDs with intensity $Dst \leq -100$ nT caused

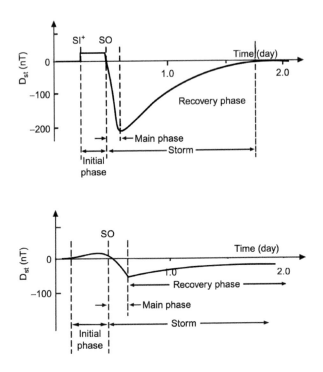

Figure 2.7: Schematic representation of GMD types: Type I on the top; Type II on the bottom (modified from [38])

by ICMEs follow the solar cycle sunspot number. There is an uneven dependence for weak-to-moderate GMDs caused mainly by CIRs [38]. Since different solar wind conditions have different intrinsic properties, they stimulate geomagnetic activity differently [50]. The frequency distribution for intense GMDs as a function of solar cycle phase and triggering event is given in [37]. The major storms at solar maximum were 100% CME-associated and 88% of them at solar minimum. Other 12% were triggered by corotating streams.

During the substorms, the longitudinal currents with a total intensity $1 - 2 \times 10^6$ A are generated. They flow over magnetic force lines and connect the magnetotail with ionospheric aurora zone. When the solar wind plasma crosses the magnetopause force lines, the protons deflect to the dusk side of the magnetotail, and electrons - to the dawn side. Current flows between these two regions in the way that most of it flows across the magnetotail and the other part along the force lines to and from the auroral oval in the ionosphere. This longitudinal current is carried by electrons that excite auroral emissions. Parallel to these primary currents, secondary currents flow. These longitudinal currents are closed in the ionosphere by forming intense auroral electrojets (western and eastern). The electrojet currents spread along the ionosphere to the subauroral and even mid-latitude regions. Magnetic field variations on the

Earth's surface from these currents can be several hours long. The heated solar wind plasma, as well as ionospheric ions with energies from 10 to 500 keV, are injected into the region of closed field lines at geocentric distances 3–7 times the Earth's radius, forming a ring current. The recovery phase is characterized by the return of geomagnetic field to an unperturbed values due to the attenuation of the ring current as a result of dissociation of energetic ions colliding with neutral hydrogen atoms in the exosphere.

## 2.3 GEOMAGNETICALLY INDUCED CURRENTS IN TECHNOLOGICAL SYSTEMS – HISTORICAL OVERVIEW OF DAMAGES AND DRAMATIC CONSEQUENCES

The Earth magnetic field characteristics are determined by the currents in the Earth's core and the currents in the Earth's magnetosphere and ionosphere. In their turn, currents in Earth's magnetosphere and ionosphere are determined by solar wind parameters. The change of magnetosphere characteristics results in the ring current appearance, which decreases magnetic field in middle and low latitudes. Simultaneously, additional ionospheric currents (east and west electrojets) appear in high latitudes. The geomagnetic field varying in space and in time induces quasi-direct electric field that drives GICs in a system of conductors. The majority of man-made critical infrastructures are no barrier to GIC distribution.

The severity of solar storm impact depends on a multitude of the factors of different nature. The detailed study and critical factors ranking as a function of their influence on GMD severity impact is given in Section 6.3.

### 2.3.1 POWER GRIDS

The first technological infrastructure that experienced solar storm impact is telegraph communication system. It is described in [51] that the spontaneous deflections in the needles of telegraph system were observed. They were especially strong whenever auroras were visible. It is worth noting the impact on critical infrastructure due to the Carrington event. The Carrington event, a solar storm that lasted from August 28 until September 2, 1859 is the strongest of its kind ever registered and one of the most famous solar storms. Due to the undeveloped infrastructure of those days, solar storms did not cause any huge technical or economic losses. Mr. Wood, superintendent of Canadian Telegraph lines, described the event as follows: "so completely were the lines under the influence of aurora borealis, that it was found impossible to communicate between the telegraph stations, and the line was closed for the night. Problems were also reported by telegraph operators in New York, Washington, Philadelphia, Vermont, Massachusetts as well as in France and England".

Six solar cycles after the Carrington event, the Railroad storm happened in May 1921. The sunspot which caused the brilliant aurora borealis on Saturday night and the worst electrical disturbance in a memory on the telegraph system was credited with an unprecedented thing at 7 o'clock yesterday morning when the entire signal and switching system of the New York Central Railroad below 125th street was put out of operation [52].

A storm in 1940 is considered to be the first solar event that affected power grids. According to [53], following disturbances were registered in total by 22 power companies in United States and Canada during the solar storm on March 24, 1940:

- voltage dips ranging up to 10 percent but generally of short duration (7 registered cases);
- tripping transformer banks by differential relay operation (5 registered cases involving 15 transformer banks);
- large increase or swings of reactive power (4 registered cases).

The disturbances of normal power system operation and equipment damage were registered as well in Norway [54]. The solar storms in 1957–1958 questioned electricity chain stakeholders by the increased impact and losses in comparison to previous events. The simultaneous trip of transformers T1 and T2 at the Port Arthur power plant caused Toronto blackout on February 10, 1958 [55]. On the same date, the TAT-1, transatlantic cable, was out of service [56].

---

**Table 2.2**

**Sequence of events: Hydro-Quebec blackout March 13, 1989**

| Time, EST | Failure | Voltage level, p.u. |
|---|---|---|
| 2:44:17 | Trip of compensator CLC 12 at Chibougamau | ∼ 0.9 |
| 2:44:19 | Trip of compensator CLC 11 at Chibougamau | |
| 2:44:33-46 | Shutdown of four static compensators at Albanel and Némiskau | |
| 2:45:16 | Trip of compensator CLC 2 at La Vérendrye | |
| 2:45:24.682 | Trip of line 7025 at Jacques-Cartier | ∼ 0.8 |
| 2:45:24.936 | Trip of line 7044 at La Vérendrye | |
| 2:45:24.948 | Trip of line 7016 at La Vérendrye | |
| 2:45:24.951 | Trip of line 7026 at Chamouchouane | ∼ 0.25 |
| 2:45:24.996 | Fault on La Grande 4 transformer T1, Phase C ∼ 1.6 | |
| 2:45:25.003 | Fault on La Grande 4 transformer T3, Phase A | ∼ 1.8 |
| 2:45:25.042 | Trip of line 7045 at Grand-Brûlé and La Vérendrye | |

---

Hydro-Québec blackout on March 13–14, 1989 was the event that shifted the status of solar storm research from curiosity to necessity by showing that the GMD effects on power grid operation were underestimated despite ongoing research. The blackout caused 9-hour outage with the total loss of 19 GW. The time between the geomagnetic field change and power outage was too small, 92 seconds, for any meaningful actions. The detailed analysis of Hydro-Québec event is given in [57] and the sequence of events that led to the failure on March 13 is presented in Table 2.2 (this table is adapted from [58]). The whole grid collapsed in 6 seconds after the events described in Table 2.2.

Trip of five 130 kV transmission lines was registered in Sweden at the same time on March 13, 1989 [38]. Up to now, Swedish power grid experienced more than 15 times erroneous relay tripping, i.e. on November 13, 1960 about 30 line circuit breaker were tripped in the 400/220/130 kV systems, many of them simultaneously, in July 1982 four transformers and 15 lines tripped causing a lot of shunt-reactor switching, on March 21, 1991 nine 220 kV lines and one transformer were tripped, and others.

Despite the dedicated research, economic losses due to the GMD in 2003 could not be prevented. This solar storm in 2003 lasted for 3 days from October 29 to October 31, and affected a wide range of critical systems: power blackout in Southern Sweden, loss of about USD 570 million ADEOS-II satellite, HF communication blackout, etc. Currently, it is the best recorded strong solar event ever. Ionospheric activity followed the three-step evolution of solar storm [28]. The most intense disturbances were observed during the main phase of the third enhancement of the GIC.

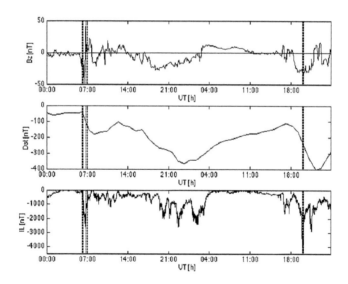

Figure 2.8: Interplanetary, magnetospheric, and ionospheric overview of October 29–30, 2003 geomagnetic storm period: (top) $z$ component of the interplanetary magnetic field measured by the ACE spacecraft, (middle) Dst index, and (bottom) local variant of the $AL$ index, $IL$ index, computed from the IMAGE magnetometer array measurements [28]

As it is seen from Fig. 2.8, the first CME arrived in the morning of October 29, (05:40 UT), while the second one in the evening October 30. Both CMEs resulted in strong geomagnetic storms Kp= 9. The geomagnetic storm intensity stayed at the level Kp=7, even when the solar wind component was positive. The minimum Dst component was equal to 401 nT [61].

The famous Halloween blackout occurred in Malmo, Sweden, on October 30, at 20:07 UT and lasted for 20 to 50 minutes affecting about 50,000 customers. The excessive amount of the third harmonics in the grid is considered to be a result of power transformers saturation caused by the GIC. The simultaneous values reached up to 300 A [124]. Dashed lines in Fig. 2.8 indicate the failure moments of Swedish power grid presented in Table 2.3 [28]. The overview of events during 2003 Halloween storm in North America is represented in Fig. 2.9.

**Table 2.3**

**Failures reported in the Swedish high-voltage power transmission system during October 29–31, 2003 storm [28]**

| Time, UT | Failure type |
|----------|--------------|
| **Tuesday, October 29, 2003** | |
| 06:11:42 | 220 kV power line from a power station was tripped and 140 MW generation was disconnected |
| 06:12:29 | 130 kV power line was tripped; the same line tripped for the second time at 07:04:10 UT |
| 06:46:04 | Trip of 400 kV transmission line Hemsjö-Karlshamn due to power imbalance caused by 300 MW power import to Sweden from Poland via the HVDC link SwePol. |
| 07:00:00 | High temperature was registered in the step-up transformer of a nuclear power plant located in Southern Sweden; temperature rise repeated several times |
| **Wednesday, October 30, 2003** | |
| 19:55:28 | Trip of 400/220 kV transformer |
| 20:03:43 | Trip of 400/130 kV transformer which caused overload in the 130 kV network |
| 20:03:44 | Trip of 130/10 kV transformer |
| 20:07:15 | Trip of a 130 kV power line in Malmo caused the blackout resulted energy undersupply for 50,000 customers |
| 20:08:00 | Trip of 130 kV line |
| 20:08:32 | Trip of 130 kV line |

Data analysis showed that the strong solar activity may also affect mid-latitude power systems. For instance, National Grid, England, experienced disturbances in solar cycles 21 and 22. Nevertheless, no widespread damage is registered in UK. Events were local in nature and only two power transformers were destroyed. The list of problems during solar storms July 14,1982, March 13–14,1989, October 19–20, 1990, November 8, 1991 [128] is given below:

- large reactive power swings on generators of ca. 50–70 MVAr per generator;
- voltage dips on the 400 and 275 kV system up to 5%;
- distribution system voltage dips of 5% on average with local dips of 20%;
- repeated generator negative sequence current alarms;
- under-voltage related tripping of certain loads with numerous starts of stand-by generators associated with telephone exchanges;
- large real and reactive power swings in interconnections between England and Scotland;
- failure of two identical power transformers (32 kV, 240 MVA) at Norwich Man and Indian Queen;
- more than average communication channel failures utilized for protection and energy management system remote terminal unit communications;
- very high levels of even harmonics currents experienced due to transformer saturation.

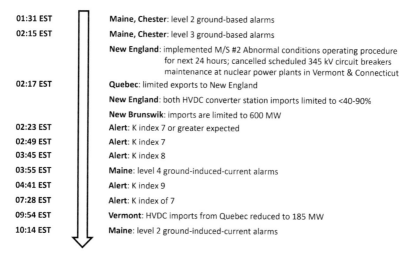

| | |
|---|---|
| 01:31 EST | **Maine, Chester**: level 2 ground-based alarms |
| 02:15 EST | **Maine, Chester**: level 3 ground-based alarms |
| | **New England**: implemented M/S #2 Abnormal conditions operating procedure for next 24 hours; cancelled scheduled 345 kV circuit breakers maintenance at nuclear power plants in Vermont & Connecticut |
| 02:17 EST | **Quebec**: limited exports to New England |
| | **New England**: both HVDC converter station imports limited to <40-90% |
| | **New Brunswik**: imports are limited to 600 MW |
| 02:23 EST | **Alert**: K index 7 or greater expected |
| 02:49 EST | **Alert**: K index 7 |
| 03:45 EST | **Alert**: K index 8 |
| 03:55 EST | **Maine**: level 4 ground-induced-current alarms |
| 04:41 EST | **Alert**: K index 9 |
| 07:28 EST | **Alert**: K index of 7 |
| 09:54 EST | **Vermont**: HVDC imports from Quebec reduced to 185 MW |
| 10:14 EST | **Maine**: level 2 ground-induced-current alarms |

Figure 2.9: Sequence of events in North America on October 30, 2003

Chinese power grids also experienced abnormal power transformer operation during 23 solar cycle. The three single phase 750 MVA transformers installed on the Shanghe substation (33.4° N, 119.2° E) were disturbed with unknown abnormal noise and severe vibration. Later on, it was concluded that the effects were caused by the GIC. Besides, the power transformer on Chifeng power plant (42.3° N, 119.0° E) experienced abnormal noise several times during the periods of high geomagnetic activity in 2003–2004 [65]. The data obtained during seven geomagnetic storms

(November 2004 – August 2005) with sampling interval of 1 second showed the GIC values up to 55.8 A (18:50 UT on November 9, 2004). The GMDs impact on power grids was registered in southern hemisphere (18–30° S) as well [26], [67], which led to power transformer outage.

Contrary to Halloween event, a solar storm on November 6–10, 2004 did not lead to the same result. The event had lower predictability in comparison to the storm on October 29–31, 2003, because the two large jumps in geomagnetic activity were caused by multiple interaction CMEs [68]. The Dst was equal to $-373$ nT during the first CME event and $-289$ nT in the latter. In general, GIC levels were not high excluding some particular areas (Fig. 2.10). The current of 6 A was registered during the storm on November 7–8, 2004, which was relatively small as compared to the one during Halloween event 42 A [69]. High GIC of 109 A was registered in Sweden, the storm on one of the sites in North America was almost 80 A. The most likely cause of current rise is not the geomagnetic situation but the particular power grid characteristics.

Compared to the previous solar cycles 22 and 23, solar cycle 24 (started in 2008) is relatively quiet. Though it did not bring any remarkable power grid outages, it provided an enormous amount of data to the scientists active in data collection and analysis.

### 2.3.2   RAILWAY AUTOMATICS

Railway is another type of technological system that experiences negative GMDs impact. The physical principle of impact is similar to the one for power grids. Research on GIC impacts on railroad has not been very extensive so far. Although solar storms have possibly affected railroads in the past, only few registered events are known. Society is less aware of GMD impact on railways than on power grids.

The first registered event happened in 1921 in New York and is called Railroad storm. Nevertheless, the detailed study of solar storm impact on railroads was started in the 1980s. [70] highlights the following effects of solar storm impact on railroads:

- communication failures;
- loss of commercial power sources;
- signaling circuits miss operation.

The change of Earth's ionosphere affects radio transmission in the high frequency and results in communication failures. However, modern communication networks designed with the use of fiber optic cables should minimize these effects. The GMDs also may decrease the quality of electric power by changing the shape of sinusoidal current to the meander.

The GMD on the night between July 13 and July 14, 1982 affected 45 km section of Swedish railway in the southern part of the country [124]. The traffic lights turned red without any obvious reason and after a while turned green back and back to red again later. Due to legal norms the normative voltage difference between the rails and over a relay is kept in the range of 3–5 V (DC). The induced voltage was large

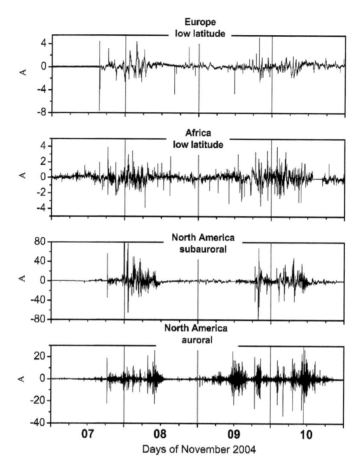

Figure 2.10: GIC measured at specific sites of four power systems, located in different regions for four disturbed days in November 2004 [68]

enough to make the voltage zero (imitating the process of train passing). Therefore, the electric field of 4–5 V/km was observed. The graphic explanation is given in Fig. 2.11.

A solar storm in March 1989 affected not only Hydro-Québec power grid but also railways in Russia [61] and Canada [71]. On March 13, 1989 at 20:45 UCT, the numerous cases of false blockages of railways tracks of Gorky railway, Russia, were registered. The most affected region is located within the coordinate range 52.1–55.13° N and 115.4–125.06° E in geomagnetic coordinates. The anomalies occurred at the times of high Kp index value.

Contrary to power grids, there are few examples of railways located in high latitudes. Russian railway is one of the examples. During the years 2000–2005, the Forecasting Center of IZMIRAN, Russia, received numerous reports about anomalies

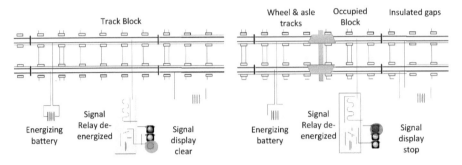

Figure 2.11: Change of signal due to presence of train wheel [63]

in the signalization system operation. The anomalies could not be explained by the usual reasons, i.e. malfunction of isolation joints or butt connectors, steel and throttle crosspieces, fall of the resistance of a ballast, breaks of a rail, etc. for instance, reports of 3,916 anomalies were received in 2004 [73]. Analysis of failure origin showed that 45% could be due to meteorologically connected reason. For other 2,176 anomalies, technical staff provided various technical reasons. Out of 2,176 anomalies 20 were with non-recognized technical reason. The example of railway automatic failure during a substorm at 18:15 UT January 19, 2003, studied by Polar Geophysical Institute is, presented in Fig. 2.12.

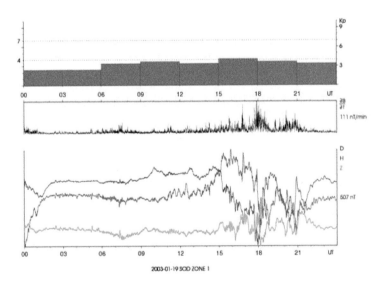

Figure 2.12: Example of railway automatic failure during substorm at 18:15 UT January 19, 2003 (Courtesy of Sakharov, Ya.)

In March and September 2004, there was no typical seasonal increase of geomagnetic activity. Instead, July and November storms dominate. Thereby, 16 out of 20 undefined anomalies were observed in the second part of the year. For yearly period, the probability of anomaly occurrence is about 5%. The probability of anomalies occurrence during disturbed days (Dst$< -50$) is about 30%. In total, 19 days with such geomagnetic activity were registered in 2004. The statistical analysis of the period 2002-2006 shows that the probability of anomaly occurrence during strong geomagnetic activity is about 60% [74].

In general, the failures in automatic system occur during high geomagnetic activity. For the period 2002–2006, failures have been registered during substorms at Oktyaborskaya railway. Hence, only 3–5% of weak magnetic disturbances cause failures in automatic. The diagram in Fig. 2.13 shows the number of failures as function of geomagnetic activity and latitude [74]. The total number of failures caused by various reasons is given in the grey color, and the number of failures caused by GMDs is represented in color. The total number of failures is marked on the top of each column. Three following geomagnetic zones were considered:

- from $68°59'$ to $67°10'$ N;
- from $67°10'$ to $64°32'$ N;
- from $64°32'$ to $60°57'$ N.

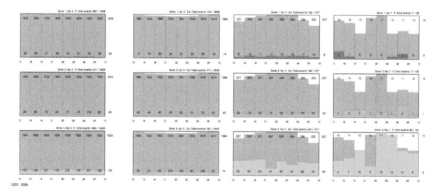

Figure 2.13: Failures occurrence in railway automatics given for three zones at different levels of Kp (Courtesy of Sakharov, Ya.)

The affected area is located in the mid-geomagnetic latitude during severe events, since the magnetosphere has stronger deformation.

The combined table of severe storms when failure occurred in Northern railways is represented in Table 2.4.

### 2.3.3 PIPELINES

Pipelines are another example of a grounded technological system that experiences negative GMD effects. Despite the fact of not registering any destructive GMD event

**Table 2.4**

**List of severe magnetic storms when failures have occurred in the automatic system of Northern railway [72]**

| Date | Max Kp | Daily $A_p$ | Min Dst,nT | Time of failures (UT) |
|---|---|---|---|---|
| March 13–14, 1989 | 9 | 246 | −589 | 21:45–01:45 |
| April 6–7, 2000 | 9− | 83 | −321 | 17:47–01:30 |
| July 15–16, 2000 | 9 | 163 | −300 | 20:05–00:25 |
| Sep. 17–18, 2000 | 8+ | 57 | −172 | 20:45–01:00 |
| March 31, 2001 | 9− | 192 | −358 | 03:05–06:30 15:05–16.06 |
| April 8, 2001 | 7 | 63 | −51 | 20:17–23:00 |
| April 11, 2001 | 8+ | 85 | −256 | 17:55–21:50 |
| Nov. 6, 2001 | 9− | 242 | −257 | From 01:52 |
| Nov. 24, 2001 | 8+ | 104 | −225 | 07:05–09:00 |
| Oct. 29, 2003 | 9 | 204 | −345 | 19:15–21:33 |
| Oct. 30, 2003 | 9 | 191 | −401 | 19:50–21:16 |
| Nov. 20, 2003 | 9− | 150 | −472 | 15:37–22:40 |
| July 22–25, 2004 | 8 | 154 | −148 | 23:10–24:00 (22/07) 12:18–12:50 (24/07) |
| Nov. 8–10, 2004 | 9− | 161 | −373 | 23:40–24:41 (8/11) 02:22–07:55 (9/11) 19:50–21:20 (10/11) |
| Jan. 21, 2005 | 8 | 66 | −105 | From 22:20 |
| May 7, 2005 | 8+ | 91 | −126 | 21:35–24:15 |
| May 14, 2005 | 8+ | 87 | −256 | 19:55–24:50 |

from a pipeline reliability perspective, the data collection in this domain rather developed compared to power grids and railways. One of the reasons is the relatively easy access for GIC measurement performing. The coating is implemented in order to protect pipelines from destruction because of corrosion. Since no coating is perfect, the cathodic protection system is applied for keeping the pipeline under the negative voltage with respect to the Earth. It is common practice to make an annual base survey to investigate the current pipe-to-soil potential (PSP) variations. Many energy pipelines are made of carbon steels types X65 and X70. The safe value for this steel is within the range 0.85–1.15 V for steel pipelines. More negative PSP may be required for mill scaled steel, when microbiologically induced corrosion is suspected, in weak acid environments, or at high operating temperatures. Less negative potentials could be acceptable for well-drained, well-aerated pipelines located in high resistivity soils or within concrete casings [75]. The PSP readings are often irregular and at times fall outside the recommended range [77]. The GIC can override a pipeline cathodic protection by changing PSP which may lead to enhanced corrosion. The high corrosion level could lead to an ecological tragedy of oil poured out of a hole in a pipeline. A set of factors determines the value of measured PSP value at a given local time:

- the geomagnetic activity [76];
- the Earth's conductivity profile [77], [78];
- the pipeline structure including bends, flanges, terminations, coating, splitting or merging one or two pipes [76], [79].

If the GIC only flows along the pipeline, it does not change the PSP value. At inhomogeneities of the pipeline or of the surrounding earth, in particular near the ends and bends of the pipeline, the GIC can flow between the pipeline and the earth [80]. A research sponsored by the American Gas Association (AGA) in 1966–1970 concluded that the effects of GIC are insignificant, both for coated, protected lines and for bare lines [81]. This conclusion cannot be implemented to modern networks because;

- the pipelines studied by the AGA had low leakage resistivity10 $k\Omega m^2$ which is 10 times lower in comparison to modern pipelines.
- all the tested pipelines were electrically short (only one longer than 65 km).
- the pipelines were located in mid-latitudes (lower than 46° N).
- even though the study took place in the period around solar maximum, following-up observations showed that it was the weakest maximum in the last 50 years.

The PSP variations due to geomagnetic activity have been reported in the system in various parts of the world. The first registered event of GMD impact on pipeline in Sweden was registered in 1991 [82]. The set of registered events afterwards is collected. The diversificated network of survey stations and the high time resolution (1 min) make the GIC recording in the Swedish pipeline system scientifically interesting [72]. During the main phase of Halloween magnetic storm, PSPs in the Czech pipeline network varied from +2 to −8 V (DRUZBA pipeline), and were much smaller on the IKL pipeline [84]. The difference originates in the properties of the coating materials. DRUZBA pipeline is coated with tar, which has lower pipe-to-soil resistance than polyethylene which is used for IKL pipeline coating. The November 2004 solar storm affected the pipelines. The voltage fluctuation was three times higher than the admissible range (−1 V to −2 V) especially during the second storm [68].

The November 2004 solar storm affected the pipelines. As seen from Fig. 2.14, the voltage fluctuation was three times higher than the admissible range (−1 V to −2 V), especially during the second storm.

The analysis of GIC distribution in Finnish pipeline network was launched in the 1980s by the means of analytical research. The Finish pipeline network is 914 km long. The pipeline is composed of main network (350 km between the eastern and western ends) and of several shorter branches. The pipeline is electrically connected to Russian pipeline network [85]. The real measurements started in 1998 using the method of two magnetometers, i.e. one just above the pipeline and another at Geophysical Observatory located in 30–40 km. Since the distance is only about the 30 km, the natural field variations are not high [86]. The recordings describe the natural variation of magnetic field for the period 1998–2005 [4]. The solar activity is characterized by two measures: the maximum absolute 10-s value and the number of 10-s values exceeding 5 A (Table 2.5).

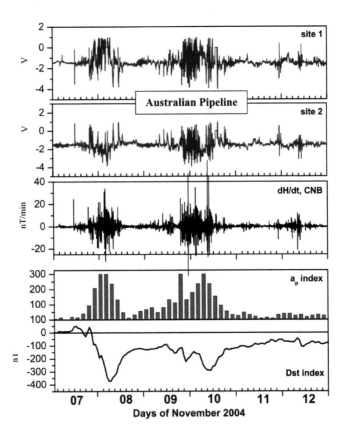

Figure 2.14: Pipeline voltages at two sites, time derivative of the horizontal component of geomagnetic field and $A_p$ and Dst indices for November 7–12, 2004 [68]

Similar to power transformers, GMDs may lead to postponing pipeline degradation. A pipeline carrying aviation fuel to Auckland International Airport failed on September 14, 2017 since mechanical digger damaged the insulating coating and scrapped the pipe itself. It is believed that the pipeline was weakened by a GMD six days earlier with Kp of 8+ at most [89]. This is not the first attempt to assess the GMD impact on pipelines in New Zealand. The research was kicked-off in 1992 by analyzing the correlation between the PSP measurements and geomagnetic field variations for a pipeline close to the town of Dannevirke in the North Island of New Zealand [90]. Both consistent phase and amplitude relationships across four pipelines in the neighboring region, Australia, between magnetometer and PSP sites were also established in [91].

**Table 2.5**

**Largest GIC days in the period 1998–2005 affected gas pipeline at Mantsala [18]**

| UT Day | Max GIC, A | GIC$^a$ | K | $\left\| \frac{dH}{dt} \right\|$ |
|--------|------------|---------|---|------|
| 2003.10.29 | 57.0 | 1,934 | 9 | 4,094 |
| 2003.10.30 | 48.8 | 878 | 9 | 2,039 |
| 2004.11.09 | 42.8 | 450 | 9 | 946 |
| 2004.11.07 | 34.8 | 205 | 9 | 412 |
| 2001.11.24 | 32.0 | 515 | 8 | 1,326 |
| 2001.11.06 | 31.6 | 579 | 9 | 1,045 |
| 2003.10.31 | 30.3 | 829 | 9 | 2,337 |
| 2000.07.15 | 30.1 | 549 | 9 | 1,195 |
| 2004.11.08 | 29.1 | 833 | 9 | 1,631 |
| 2003.10.14 | 28.7 | 79 | 9 | 124 |

$^a$ The number of GIC value exceeding 5 A

In conclusion, we re-stress that scientists agree on the assumption that the Carrington event is the strongest witnessed event, though the full picture is depicted with considerable uncertainties. For example, inferences of Dst range from −850 nT to more than −1,700 nT [92]. Another degree of uncertainty is the level of impact in case of its recurrence. In 1859, the GMD impacted only telegraph system. Modern and future technologies about which we are conscious and which we consider to be more sensitive to GMD was not yet developed. Ice core records have been used to determine the spectral slope and the influence of very large SPEs [93]. However, many ice core data do not show a nitrate spike related to the Carrington event [94]. Moreover, CME arrival time is estimated as 16.5 hours making CME one of the fastest [92]. Modern geophysics classify CME on July 23, 2012 as a reasonable worst-case scenario. Thanks to observational technology improvement, it is recorded and described. We are able to obtain phenomenal measurements of the event and conclude that, had it hit the Earth, it would likely have caused catastrophic damage [95]. Detailed information on economic loss evaluation caused by severe GMDs is given in Section 6.5.

## 2.4  GROUND EFFECTS OF GEOMAGNETIC DISTURBANCE MODELING

In general, the task of ground effects of GMD modeling can be divided into four steps (Fig. 2.15):

1. **Geophysical step**, which includes the calculation of geoelectric field using the data from the closet geomagnetic observatory and the data of surface impedance (Section 2.4.2).
2. **Electrotechnical step**, which is devoted to GIC determination in the power grid with the given characteristics (Section 2.4.3).
3. **System step**, within which the estimation of negative effects from GIC on power grid operation is done (Chapter 4).
4. **Economic loss analysis and mitigation strategies planning**, which are discussed in Chapters 5 and 6.

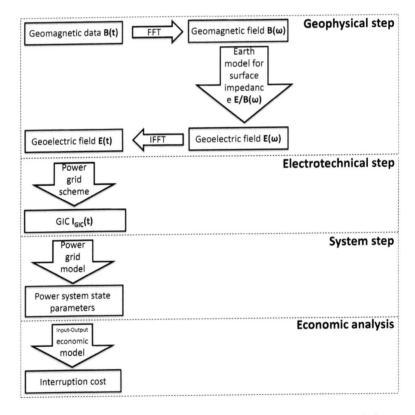

Figure 2.15: Algorithm of ground effects of space weather simulation

## 2.4.1 GEOMAGNETIC DATA

Geomagnetic field variations are continuously recorded around the globe. This data is used to characterize GMDs strengths and duration. Two completely different patterns of GMDs can be identified. The characteristic feature of GMD in high geomagnetic latitudes is large and rapid temporal variations together with the strong spatial inhomogeneity in the rate of change of the primary geomagnetic field [96]. The strong spatial inhomogeneity of the horizontal magnetic field rate of change prior (20:06:30 UT) to the power grid blackout experienced in Sweden at 20:07 UT is shown in Fig. 2.16 [28]. It is important to note that the ground magnetic field is relatively smooth at 20:07 UT. In contrast, GMDs at low geomagnetic latitudes have small and smooth variations $<\pm500$ nT. Between these two regions ($48°$–$57°$ latitudes) an intermediate type can be recognized.

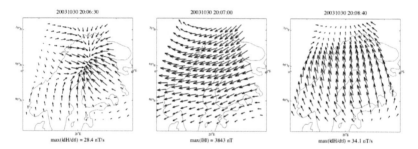

Figure 2.16: From left to right: the time derivative of the horizontal ground magnetic field at 20:06:30 UT; equivalent ground currents formed by rotating the ground magnetic field at 20:07:00 UT; the time derivative of the horizontal ground magnetic field at 20:08:40 UT [98]

Observation results for St. Patrick's Day storm on March 17–18, 2015 showed the same pattern [99]. The character of magnetic field variations is latitude dependent. The magnetic field variations in the high-latitude were observed 10 hours after SSC and they were the largest in their nature compared to those in the lower latitudes. Magnetic field variations at the mid-latitude occurred at the moment of the storm's commencement predominantly on the dayside. Pattern of magnetic field variations at the equatorial latitudes was the mixture of two above, i.e. the variations were observed both at the moment of the storm's commencement and 10 hours later.

The input for geoelectric field modeling is the data on spatial and temporal geomagnetic pattern. Geomagnetic datasets are derived from ground-based geomagnetic field measurements. The epoch of magnetic field measurement was started in the early nineteenth century in response to the influence of Alexander von Humboldt and Carl Friedrich Gauss. Some of those stations still run. Despite intense development of other technological sectors, many stations were still operating in a "classical" mode with analogue recording and manual data processing in the 1980s. Two types of stations exist: magnetic observatories and variometer stations. Magnetic observatories are designed in support of continuous and accurate measurements of the local geomagnetic field over a long duration of time [100]. In contrast, variometer stations are simpler in operation. Their data is rigorously calibrated for absolute accuracy. Variometer deployment helps to fill the gaps between the sparse distribution of magnetic observatories. Modern observatory has a fluxgate magnetometer, which gives vectorial data conventionally expressed in terms of either the Cartesian components (X [*north*], Y [*east*], and Z [*down*]) or the horizontal polar components (horizontal intensity $H = \left[X^2 + Y^2\right]^{\frac{1}{2}}$, declination $D = arctan\left[\frac{Y}{X}\right]$, and Z [*down*]). For achieving higher measurement accuracy, modern observatories have a proton precession magnetometer that measures the total absolute field intensity $F = \left(X^2 + Y^2 + Z^2\right)^{\frac{1}{2}}$.

The International Real-Time Magnetic Observatory Network (INTERMAGNET) was founded in late 1980s in order to prompt the operation of magnetic observatories according to modern standards. INTERMAGNET institutes have cooperatively developed infrastructure for data exchange and management as well as methods for data processing and checking [101]. To reliably produce a long-period geomagnetic time series, an observatory must operate under carefully controlled conditions [102]. It was essential to convince the international observatory community that adopting modern technologies and data processing would not comprise the data quality required for secular variation studies. This has been clearly demonstrated. As a result about half of world's observatories were INTERMAGNET members in 2001 [103]. Currently, INTERMAGNET consists of ca. 170 observatories operated by government, academic and commercial institutes. An interface between individual observatories and data users is real-time data collection centers, designated Geomagnetic Information Nodes (GINs). They are connected to the INTERMAGNET observatories by satellite, computer and telephone networks. There are GINs in North America (USGS, Golden and GSC, Ottawa), Europe (BGS, Edinburgh and IPGP, Paris) and Japan (Kyoto and Hiraiso). The Kyoto GIN is run by the World Data Center for Geomagnetism at Kyoto University. The GIN in Hiraiso is operated by the Hiraiso Solar Terrestrial Research Center, Communications Research Laboratory. The picture of GINs and corresponding satellites is presented in Fig. 2.17.

INTERMAGNET originally focused on producing 1-minute average magnetic data. Around 120 observatories produce and routinely report digital data with an acquisition cadence of one minute or better. The remaining 50 or so observatories use older, analog systems or report their data only years after acquisition [104]. Two types of vector data are produced: preliminary data – unprocessed records of the geomagnetic vector acquired by a fluxgate magnetometer which should be available to users within 72 hours; definitive data – processed data which is the combination of fluxgate date with auxiliary measurements of absolute field direction and intensity to correct for fluxgate orientation and baseline drift [102].

If magnetic data is only available from one observatory in the area of a particular system, then using data for calculations across the whole power system involves the implicit assumption that the magnetic field variations are spatially uniform across the power grid [105]. In case data from two magnetic observatories located in the opposite ends of the power grid exist, then the magnetic field variations at sites across the power system can be approximated by linear interpolation from the observatories

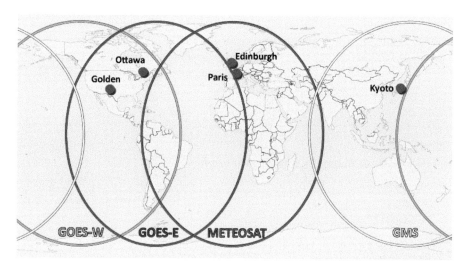

Figure 2.17: Geomagnetic information nodes (Courtesy of INTERMAGNET: www.intermagnet.org)

recordings [106]. Nevertheless, uniform or linear approximation is not always valid, since the magnetic field inhomogeneity exists. Higher modeling accuracy can be achieved by the use of more magnetic recordings site, then an interpolation scheme to provide magnetic values across the power grid can be used. If the area is large enough, the Earth's surface cannot be assumed planar. Spherical schemes such as spherical elementary current method [107] and spherical cap harmonic analysis [108] should be used.

Measured geomagnetic data is used as an input for GMD models for reproducing observed features in the signal of interest. Inaccurate model estimates and misleading errors may lead to poor and costly decisions by the end users. The $dB/dt$ value is used for the validation based on the argument that the time derivative of the ground magnetic field is an indicator for the level of geomagnetically induced electric field or geoelectric field on the Earth's surface [109]. The model performance is evaluated by comparing the observed versus predicted ground $dB/dt$ as in Eq. 2.4:

$$dB/dt = \sqrt{(dB_{\text{north}}/dt)^2 + (dB_{\text{east}}/dt)^2},$$  (2.4)

where $B_{\text{north}}$ and $B_{\text{east}}$ are two horizontal components of the magnetic field (geomagnetic dipole coordinates).

The standard introduced by [110] has several limitations. In order to overcome them, the Ground Magnetic Perturbation working team of the International Forum for Space Weather Capabilities Assessment was established [111]. Recommendations on model validation expansion include:

- Consideration of the new geomagnetic events. Originally, four events were chosen (first four lines in Table 2.6) [112]. Later two more events were added (lines 5–6 in Table 2.6) [110]. The working group included two more events and considered five others (lines 7–8 and 9–13, respectively, in Table 2.6). For each auroral electrojet the start time, duration over-which data-model comparison ought to be made, maximum F10.7 solar flux, Kp, AE and minimum SYM-H values are given.

**Table 2.6**

**List of events for magnetic model validation**

| no. | Event start | Extent, h | $F10.7$ | Kp | AE, nT | SYM-H, nT |
|-----|-------------|-----------|---------|-----|--------|-----------|
| 1 | 2003.20.29 06:00UT | 24 | 275.4 | $9^0$ | 4,056 | −391 |
| 2 | 2006.12.14 12:00UT | 36 | 90.5 | $8^+$ | 2,284 | −211 |
| 3 | 2001.08.31 00:00UT | 24 | 203.0 | $4^+$ | 959 | −46 |
| 4 | 2005.08.31 10:00UT | 26 | 96.0 | $7^0$ | 2,063 | −119 |
| 5 | 2010.04.05 00:00UT | 24 | 113.0 | $8^-$ | 2,655 | −67 |
| 6 | 2011.08.05 09:00UT | 36 | 90.5 | $8^-$ | 2,611 | −126 |
| 7 | 2015.03.17 02:00UT | 34 | 116.0 | $8^-$ | 2,298 | −234 |
| 8 | 2004.07.22 06:00UT | 162 | 178.4 | $9^-$ | 3,632 | −208 |
| 9 | 2004.11.07 00:00UT | 60 | 138.1 | $9^-$ | 3,360 | −394 |
| 10 | 2001.03.30 12:00UT | 48 | 257.2 | $9^-$ | 2,407 | −437 |
| 11 | 2013.03.17 00:00UT | 48 | 124.5 | $7^-$ | 2,689 | −132 |
| 12 | 2000.04.06 12:00UT | 48 | 178.1 | $9^-$ | 2,481 | −320 |
| 13 | 2005.05.15 00:00UT | 24 | 105.2 | $8^+$ | 2,051 | −305 |

- Increase coverage and resolution of observations. The current sampling rate is 60 s, because most of magnetometer stations release 1 min data. While a 60 s sampling rate captures most GIC-pertinent fluctuations, a 1 s resolution is optimal [79]. The lower time resolution observations also limit the quality of the numerical derivative of $\Delta B$ [114]. Currently, 400 stations exist in

Northern hemisphere, while only 31 of which report data at a 1 s frequency.

- Transition from traditional per-station to regional analysis. Magnetic observatories were segregated into two groups based on latitude, but did not provide information about model performance as a function of magnetic local time (MLT) in [110]. A $dB/dt$ peak has to be predicted correctly both temporarily and geographically. Simple MLT binning method is recommended for improving the accuracy. A set of virtual magnetometers spaced at $5°$ latitude and longitude intervals across the globe is included as a part of a model. It will result in creation of contingencies tables and metrics as a function of MLT quadrants instead of a per-station basis. Such MLT binning should be implemented alongside current latitudinal segregation.
- Segregation by activity type means assessing different activity types separately. Currently, GMDs are considered as a net effect of many subsequent events. It is believed that it is feasible to consider three activity types: SSCs, substorm expansions, and ring current intensifications.
- List of metric adjustment. The process to quantify the data-model comparison based on binary event analysis was introduced by [115]. Following list of metrics is currently used:

1. The probability of detection (POD) – the fraction of observed threshold crossing predicted by the motel, also called hit rate and is defined as in Eq. 2.5:

$$POD = \frac{a}{a+c},\qquad(2.5)$$

where $a$ is the number of hits, $b$ is the number of false positives, and $c$ is the number of misses.

2. The probability of false detection (POFD) – the probability of an event being correctly predicted by the model, given that an event occurred defined as in Eq. 2.6:

$$POFD = \frac{b}{b+d},\qquad(2.6)$$

where $d$ is the number of true negatives.
Smaller values of POFD stays for better model performance.

3. Heidke Skill Score (HSS) – the accuracy relative to a reference model (Eq. 2.7).

$$HSS = \frac{2\,(ad - bc)}{(a+c)\,(c+d) + (a+b)\,(b+d)}\qquad(2.7)$$

4. The frequency bias (FB) defined as the ratio of threshold crossing forecasts (including false positives) to the observed crossing forecasts (Eq. 2.8). It is a newly proposed metric.

$$FB = \frac{a+b}{a+c}\qquad(2.8)$$

The International Living With a Star Steering Committee of the Committee on Space Research together with Inter-programme Coordination Team on Space Weather (ICTSW) of the World Meteorological Organization (WMO) recommend prompt access to data from ground-based magnetometers [167]. The United States National Space Weather Action Plan identifies real-time ground-based magnetometers as an observational capability that should be expanded [137]. Nevertheless, opportunities in enhancing ground-based geomagnetic monitoring are limited by institute-to-institute differences in capabilities, traditions, interests, fundings, priorities, policies, and cultures.

### 2.4.2   GEOELECTRIC FIELD MODELING

The horizontal geoelectric field occurring on the Earth's surface during a GMD is the key quantity for the GIC in a ground-based technological system [118]. The induced geoelectric field observed on the surface is independent of the technological system and depends on magnetosphere-ionosphere current system, which in turn is dependent on space weather conditions, and on electromagnetic induction that is determined by the Earth's geology [119]. Each event is unique, hence, there is no typical geoelectric field pattern. Although the basic physical principles behind the phenomena stay unchanged.

A simple way to estimate the geoelectric field is to assume that the primary space contribution to the geomagnetic variation and to the geoelectric field on the Earth's surface is a downward-propagating plane wave and that the Earth is uniform or layered [120]. This is so-called plane-wave method introduced by [121]. In this case, the electric and magnetic fields are horizontal and spatially constant on the Earth's surface. If the Earth is uniform a time domain integral relation can be obtained [122], which easily allows for the use of magnetic observatory data to derive GIC statistics [123].

The basic magnetotelluric equation correlates a horizontal electric field component $E_y$ with a perpendicular horizontal magnetic field component $B_x$ (Eq. 2.9):

$$E_y = -\sqrt{\frac{\omega}{\mu_0 \sigma}} e^{j\frac{\pi}{4}} B_x, \tag{2.9}$$

where $\mu_0$ is permeability in vacuum. Equation 2.9 presumes that displacement current is neglected, since GMD frequency is ultra-low. Using an inverse Fourier transform, time domain ratio between the electric and magnetic fields can be obtained as in Eq. 2.10:

$$E(t) = -\frac{1}{\sqrt{\pi \mu_0 \sigma}} \int_0^\infty \frac{g(t-u)}{\sqrt{u}} du, \tag{2.10}$$

It is seen that $E(t)$ is affected by past values of $g(t)$, but their weight is decreasing with time. In case the multi-layered Earth is considered, Eq. 2.9 transforms to Eq. 2.11. Conductivity $\sigma$ gets the role of an apparent conductivity dependent on the frequency reflecting the penetration depth of the electric and magnetic fields into the

Earth.

$$E_y = -\frac{Z}{\mu_o} B_x,$$                     (2.11)

The validity of the plane-wave model is questionable for the GMD modeling at high latitudes due to the vicinity of localized ionospheric primary sources. More realistic ionospheric electrojets were applied by [124] and [125]. The simplest ionospheric auroral electrojet current model is an infinitely long horizontal line current at the height $h$ (Fig. 2.18). Using the standard Cartesian $xyz$ coordinate system with the $xy$ plane at the Earth's surface and the $z$ axis pointing downwards, and letting the current to be oriented parallel to the $y$ axis (unit vector $e_y$) and lying in the $yz$ plane, the current density is calculated from Eq. 2.12 [120]:

$$j(x,z) = J\delta(x)\,\delta(z+h)\,e_y,$$                     (2.12)

where $\delta$ denotes to the Dirac delta function and $J$ gives the amplitude and phase of the current.

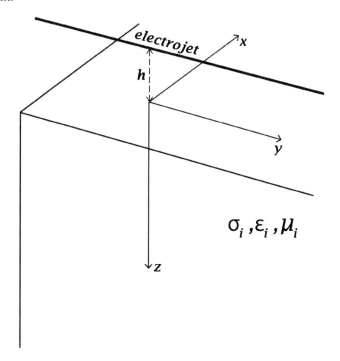

Figure 2.18: The ionospheric auroral electrojet current model

The primary electromagnetic field created by the current (Eq. 2.12) only has $y$-component of electric field and $x$- and $z$-components of magnetic field. Expression for the fields are obtained by first determining the primary fields, then solving Maxwell's equations in the air and in the Earth layers and finally using boundary conditions [119]. At the earth's surface the fields components are defined in Eqs. 2.13, 2.14, 2.15 [126]:

$$E_y(x,z,t) = -\frac{i\omega\mu_0 J}{\pi}e^{i\omega t}\int_0^\infty \frac{e^{-bh}}{b+\frac{1}{p}}\cos bx\, db \tag{2.13}$$

$$B_x(x,z,t) = \frac{\mu_0 J}{\pi}e^{i\omega t}\int_0^\infty \frac{\frac{1}{p}e^{-bh}}{b+\frac{1}{p}}\cos bx\, db \tag{2.14}$$

$$B_z(x,z,t) = -\frac{\mu_0 J}{\pi}e^{i\omega t}\int_0^\infty \frac{be^{-bh}}{b+\frac{1}{p}}\sin bx\, db \tag{2.15}$$

where $p$ is the complex skin depth $p = p(b) = \frac{Z(b)}{i\omega\mu_0}$, $Z(b)$ is the surface impedance depending on $\omega$ and particular Earth structure, and $b$ is horizontal wave number related to the $x$ coordinate perpendicular to the current.

The electric and magnetic fields at the Earth's surface produced by a primary sheet current distribution $J\beta(x)$ at the height $h$ and parallel to the $y$ axis are directly obtained by superposing line current fields, i.e. by applying Eqs. 2.13–2.15 to each individual line current $J\beta\left(x'\right)dx'$ lying at $x = x'$ and by integrating over $x'$ [126].

The real auroral electrojet has a finite length. The numerical comparison between electromagnetic fields at the surface of the layered Earth due to an infinitely long line current with those produced by a finite line current, in which upward or downward vertical (field-aligned) currents at the ends ensure current continuity, was done by [127]. The latter current system is called U-shaped current and can be described as in Eq. 2.16:

$$J_y(x,y,t) = Je^{i\omega t}\delta(x)\left(\theta\left(y+\frac{L}{2}\right) - \theta\left(y-\frac{L}{2}\right)\right) \tag{2.16}$$

where $L$ is the current length.

Complex image method (CIM) allows us to minimize the computation complexity of the secondary (Earth) contributions. The CIM was introduced in Earth's electromagnetic induction studies in the 1970s [128]. In CIM, the Earth's contributions are approximated by assuming a perfect conductor at a certain depth. In other words, the secondary field is generated by a mirror image of the primary ionospheric-magnetospheric source (Fig. 2.19). The electromagnetic field is then given by a source and imaginary currents, and their distance from the location on the surface $x$ is shown in Fig. 2.19. The depth of a perfect conductor is complex and determined by $p$ defined above (b = 0). CIM has two restrictions:

- the field-aligned currents are assumed vertical. This assumption gives very small errors in ionospheric electrodynamic studies even at high latitudes [129];
- Earth must have a layered conductivity structure.

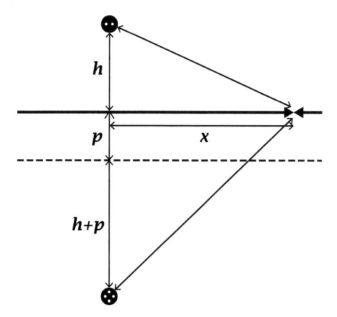

Figure 2.19: Complex image method

The CIM is only applicable in the frequency domain, hence, Fourier transform is required between time and frequency domains. The practical procedure to determine the magnetic and horizontal electric fields at the Earth's surface based on CIM is the following [120]:

- the east-west and north-south currents are given at each ionospheric grid element as functions of time. It is reasonable to consider a time interval between 10 s and 1 minute.
- Fourier transform is performed from the time to the frequency domain;
- the plane wave complex skin depth is determined for each frequency;
- the magnetic and horizontal electric fields are calculated from analytical CIM formulas at pre-defined sites at the Earth's surface for each frequency;
- all U-shaped current elements are handled in the same way, and the total field is the sum of the fields caused by induced elements;
- the time domain fields are obtained by an inverse Fourier transform.

Modeling conductivity structure of the Earth is at least as complicated as the magnetic structure of CMEs, adding another set of uncertainties to this already thorny problem. Earth conductivity varies in all directions, but the most significant change is with increasing depth within the Earth (Fig. 2.20, adapted from [134]). The resistivity decreases within the mantle and core due to the rising temperature and pressure with increasing depth. This change in the electromagnetic structure of the Earth's surface requires the adoption of the modeling techniques. The simplicity of one-dimensional (1-D) model compared to others made it the most used method for routine GIC modeling, though limitations in data affect the calculation accuracy. Therefore, the 1-D models should be interpreted only as effective approximations that allow reproducing geophysics conditions satisfactorily [130]. Nevertheless, the 1-D model can be used in inland areas where the lateral changes of conductivity can be neglected [131]. In contrast, the 1-D model can not capture the realistic picture of the geoelectric field amplification in coastal regions due to lateral conductivity gradients [132]. In case of limited availability of more advanced conductivity models, a piecewise approach can be used to provide an approximate way of taking into account lateral changes in the conductivity structure [133].

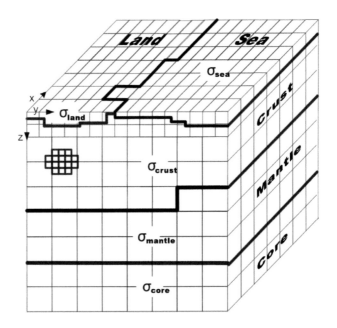

Figure 2.20: Example of the Earth structure

There are several approaches to model with the different degree of accuracy:

- **1-D models**
  - **Homogeneous model**, which is the simplest model. The conductivity is considered to be a certain value $\sigma$. In this case, there is no reflected wave traveling back up and the electric and magnetic fields can be presented as in Eq. 2.17.

$$E(x) = Ce^{-kz} \qquad (2.17a)$$

$$H(x) = \frac{1}{Z}Ce^{-kz} \qquad (2.17b)$$

where $Z$ is the characteristic impedance $Z = \sqrt{\frac{i\omega\mu_o}{\sigma}}$.

☐ **N-layered model**, which is a 1-D model consisting of a half-space $z>0$ divided into N layers with the $n$ layer having conductivity $\sigma_n$ and thickness $d_n = z_{n+1} - z_n$. The example of N-layer model is given in Fig. 2.20. The displacement current can be neglected, since GMD frequencies are very low. Then Maxwell's equations can be solved by transmission line analogy after taking the boundary conditions as in [135]. The general expression of the impedance for each layer is in Eq. 2.18:

$$Z_n = Z_{0n} \frac{1 - L_{n+1}e^{-2k_nd_n}}{1 + L_{n+1}e^{-2k_nd_n}}, \tag{2.18}$$

where $k_n$ is the propagation constant of $n$ layer $k = \sqrt{j\omega\mu_0\sigma}$, $Z_{0n}$, the impedance of the bottom layer $Z_{0n} = \frac{i\omega\mu_o}{k_n}$, $L_{n+1} = \frac{Z_{0n}-Z_{n+1}}{Z_{0n}+Z_{n+1}}$.
The propagation constants and, hence, the resulting impedances are the functions of frequencies, so the calculation has to be repeated for each frequency. The simplest version is the two-layer model, which can be an approximation of the seawater-seafloor or crust-upper mantle structures [136].

- **Two-dimensional (2-D) models**, which may be used for approximation of many geophysical structures. For instance, coast effect can be modeled using 2-D model. The three-segment model was developed for studying anomalous magnetic variations in coastal region associated with electric currents induced in the highly conducting oceans [137]. The theory was extended to a multi-segment model by [138].
- **Three-dimensional (3-D) conductivity models**, which provide the best approximation for real earth conductivity structures.

Resistivity values for multilayered models can be obtained from a variety of geophysical techniques. Near surface resistivity values are got by measuring rock samples from boreholes, which provide information for the regions immediately adjacent to the borehole and down to several kilometers. Resistivity surveys can provide information on conductivity structure down to depths of few hundred meters. Resistivity surveys measure the distribution of electrical potential in the ground around a current-carrying electrode. Information accuracy depends on the electrode spacing. Information about deeper depths can be obtained using passive techniques that utilize the natural geomagnetic field variations as the test signal. Two techniques are used:

- geomagnetic depth sounding which is based on measurements of the vertical magnetic field variations to show the presence of current concentrations. It allows us to identify higher conductivity zones within the earth.
- magnetotelluric surveys that combine the measured horizontal magnetic field variations with measurements of the surface electric field. The magnetotelluric technique has been matured in the last decades. The empirical impedance tensors obtained from magnetotelluric surveys can be used

as a way to "bypass" the need for specific conductivity models. Magnetotelluric measurements of geomagnetic and geoelectric field variations obtained from temporary deployments of sensors at individual geographic locations [139] can be expressed as empirical impedance tensors [140], and
these, in turn, can be inverted [141] to obtain models of Earth conductivity
[142]. Mathematically this approach is expressed as in Eq. 2.19:

$$\mathbf{E}(\omega) = \frac{1}{\mu} \begin{bmatrix} Z_{xx} & Z_{xy} \\ Z_{yx} & Z_{yy} \end{bmatrix} * \mathbf{B}(\omega) \qquad (2.19)$$

where $\mathbf{E}$ is the horizontal magnetic field, $\mathbf{B}$ is the horizontal ground magnetic field, $Z$ is the surface impedance tensor components. Following the
standard magnetotelluric method, temporary instrument installations can be
deployed to measure local $\mathbf{E}$ and $\mathbf{B}$ from which the impedance tensors can
be obtained [143]. All quantities are expressed in the frequency domain in
Eq. 2.19. Modern studies involve multistation surveys, producing data that
are precessed using sophisticated inversion software to obtain 2-D and 3-D
models [144, 145].

The evolution of Earth's conductivity model in Hydro-Québec region and the improved accuracy in geoelectric field modeling is presented in [146]. The resistivity
parameter for the first model is chosen as an average of resistivities within the crust of
the earth (100–10,000 Ω-m). The first attempt to create N-layer model was performed
in 1994 [147]. The surface crustal layer consists of very old igneous rocks with high
resistivity (40,000 Ω-m), followed by reduced resistivity of a mantel (4,000 Ω-m),
and a core (700 Ω-m), and a bottom half-space (1 Ω-m). The model made by Ferguson and Odwar based on their extensive geophysical literature review was presented
in 1997 (Model 3). The advanced and the current model consisting of 6 layers with
the adjusted resistivity values was presented one year later [148]. The geoelectric
field perturbations for Hydro-Québec event was modeled using four Earth conductivity models and the magnetic field data registered at Ottawa (OTT) and Poste-de-
la-baleine (PBQ) observatories. The electric field amplitudes for different scenarios
are represented in [146].

Since GMDs can have a global footprint beyond the borders of a specific region,
global conductivity models are deployed. The global scale response occurs primarily in the deep mantle. Magnetic satellite missions such as CHAMP (2001–2010)
and SAC-C (2002–2013) are data sources to derive deep (up to 1,500 km) radial
conductivity models [149]. It should be remembered that the real earth conductivity structure is not homogeneous, but is layered. The conductivity changes laterally.
Elaborated conductance (conductivity-thickness product) maps can be used to model
the ocean effect of geomagnetic storms [150].

One of the most stressing challenges in geoelectric field modeling is defining its
extremes. Extreme geoelectric scenario needs to specify both the spatial and temporal
evolution of the geoelectric field. [30] employed the plane wave method to model
extreme events, e.g. Carrington-type event also employed the plane wave method to
study the surface geoelectric field response based on a hypothetical simulation of

an extreme Carrington-type event. The modeled fields captured the global response pattern. Therefore, indications thus far suggest that the approximation can be applied also to extreme cases, but further investigations are still required.

### 2.4.3 GEOMAGNETICALLY INDUCED CURRENTS MODELING IN POWER GRIDS – DISCRETE SYSTEM ON THE CONTINUOUS HALF-SPACE

Secure and reliable power grid operation within admissible quality limits can only be ensured through high quality estimates of contingencies landscape over the time. The GIC neutral current monitoring systems give information "at-the-instant" of the event which is mainly useful for post-event analysis (see Chapters 3 and 5). Therefore, the need for the GIC modeling is predetermined by the need to develop mitigation plans for power grids. In recognition to the rapidly growing interest in the GIC assessment, the first NASA Living With a Star (ILWS) Institute Working Group that specifically targets GIC issue was established in 2014. The group was tasked to:

1. identify, advance and address the open scientific and engineering questions pertaining to GIC;
2. advance predictive modeling of GIC;
3. advocate and act as a catalyst to identify resources for addressing the multidisciplinary topic of GIC.

Faraday's law is used to explain how a GMD actually drives GIC, which correlates the electric field integrated round the loop formed by a power line, grounding wires, and return path along the Earth's surface to the rate of change of the magnetic flux through the loop (Fig. 2.21). The zero integral round the loop formed by the transmission line and the Earth's surface shows that the electric field in the transmission line has the same magnitude and direction as the electric field along the Earth's surface.

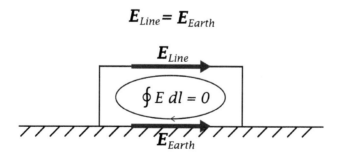

$$E_{Line} = E_{Earth}$$

Figure 2.21: Geomagnetic induction in power systems

One of the most important characteristics of the GIC is its instant magnitude. Many factors determine the GIC induced at any particular time and place. The GIC's inconsistency in time at different locations is [84]. It is shown that moving from high to low latitude decreases the peak GIC. In order words, high GIC may occur in high-latitude regions with a resistive earth structure, while small GIC is expected in low latitudes with a conductive earth structure. The correlation is not evident, since a decrease of the Earth's conductivity increases the grounding resistance, which results in smaller GIC amplitudes. The relation is investigated in greater details in [152] by considering a single power line and assuming a uniform Earth and a plane wave electromagnetic field [122]. It is presented that for practical resistance values of a grid GIC amplitudes decrease with an increasing conductivity. Nevertheless, for a critical line length the influence of grounding resistance diminishes. In general, as shown in Eq. 2.20, the GIC amplitude is proportional to:

$$I_{GIC} = \frac{E}{R_{node} + R_{TL}}, \qquad (2.20)$$

where $E$ is the electric field value across the transmission line, $R_{node}$ is the node resistance, $R_{TL}$ is the transmission line resistance.

Line length determines which part of denominator in Eq. 2.20 impacts the GIC amplitude. If the line length is less than 50 km, the GIC values are limited by the node resistance. Moreover, nodes located within 20 km distance from each other are considered as a single node. Opposite behavior is observed when the line length is larger than 300 km, i.e. long transmission line according to electrical engineering description. In this case, Eq. 2.20 transforms to Eq. 2.21:

$$I_{GICmax} = \frac{E}{r}, \qquad (2.21)$$

where $r$ is the transmission line resistance per unit length. In other words, the maximum GIC amplitude value does not depend on the line length and is determined by the specific transmission line resistivity.

The simplified grid consisting of five substations (Fig. 2.22) with various set of parameters was the scope of a study in [153]. In case all the resistance values are equal ($r_1 = r_2 = r_3 = r_4$, $R_A = R_B = R_C = R_D = R_E$), the currents are symmetrical $I_1 = I_2 = I_3 = I_4$, but GIC in the middle of the network is slightly larger than on the edges $I_2, I_3 > I_1, I_4$. It is the representation of the so-called edge effect. The dependence of GIC on the line length at the central substation is different from the ones on the edge substations as the amplitudes are much smaller and may approach zero with increasing line length. The reason for this is that with the increase of the line length, the GIC approaches an upper limit value in accordance to Eq. 2.21 [83]. In case of an asymmetrical network analysis, the largest GIC values are occurred at the edges of the network independently of whether the longer lines are at the edge. In the real power grids, "edge nodes" may occur in the middle of the physical grid model due to some geographic factors such as mountains, lakes, etc. In other words, it happens when no line goes though the node. These substations may be described as a "quasi edge effect". The same effect can be reached by calculation scheme misrepresentation when neighboring power grids are not correctly presented. It is shown in [169] ignoring the neighboring network significantly decreases the accuracy of obtained GIC values.

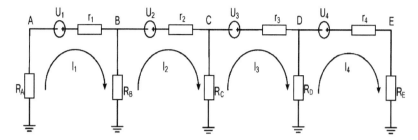

Figure 2.22: Five substation power network scheme

For modeling GIC in the power grids, the knowledge of resistances of the transmission lines, power transformers and grounding is required. These resistances are then used to construct a network model which is used with the electric fields as input to give the GIC in each branch of the network [155]. It is important to note that transmission line parameters have to be calculated by considering the sphere structure of the Earth. The Earth is an ellipsoid with a smaller radius at the pole than at the equator, hence, the precise values depend on the Earth model used. The WGS84 model is recommended by [131]. The real power grids are three-phased though one-phase line representation is used for GIC modeling. The grid is designed in the way to assure the symmetry of different phases parameters. In this case each phase of power grid will experience the same level of GIC effects. Thus the given power grid parameters have to be adjusted. This is often done by presenting the real power grid as an equivalent grid. The transmission line and transformer resistance of the equivalent grid are then calculated by diving by 3 the real values. It is important to note that the resistance of a single phase power transformer should not be divided by 3. The grounding resistance of the equivalent grid remains the same as for the real grid, since currents from all three phases flow through the same grounding resistance. Nevertheless, the calculated GIC amplitude in the grounded neutral should be divided by 3 for representing GIC per phase.

GIC calculations can be made using either the Mesh Impedance Matrix method, the Nodal Admittance Matrix method or Lehtinen-Pirjola method. In the first case, the power grid is considered as a mesh of loops with geovoltages in series in each loop, while the grid is represented as a set of earthed nodes connected together, and the geovoltages are replaced by an equipment current source in nodal admittance matrix method. In other words, each resistive branch is replaced by its corresponding admittance value. The GIC calculation using the mesh impedance matrix method is inconvenient in large real power grids. Hence, only nodal admittance matrix and Lehtinen-Pirjola methods are described further in details.

Let us consider a discretely earthed system with N nodes. The First Kirchhoff's law says that the sum of currents in each node is equal to zero. It can be used to give a set of equations relating the current and nodal voltages to the current sources at each node (Eq. 2.22)

$$j_k = i_k + v_k \sum_{n=1}^{N} y_{nk} - \sum_{n=1}^{N} v_n y_{nk}, \tag{2.22}$$

where $i_k$ is a current to ground.

In the matrix form, Eq. 2.22 can be written as Eq. 2.23

$$[\mathbf{J}] = [\mathbf{Y}][\mathbf{V}], \tag{2.23}$$

where the column matrix $[\mathbf{J}]$ is the current source vector with elements given by summing the current sources directed into each node $J_k = \sum_{n=1}^{N} j_{nk}$; $[\mathbf{Y}]$ is the admittance matrix.

The diagonal elements of admittance matrix are the sum of the admittances of all paths connected to node, and the off-diagonal elements are the negatives of the admittances between nodes $i$ and $j$. The network admittance matrix $Y_{n,ij}$ is defined as in Eq. 2.24

$$Y_{n,ij} = -\frac{1}{R_{ij}^n}(i \neq j); \ Y_{n,ij} = \sum_{k=1, k \neq i}^{N} \frac{1}{R_{ik}^n}(i = j), \tag{2.24}$$

where $R_{ij}^n$ denotes the conductor resistance between nodes $i$ and $j$. If there is no conductor between nodes $i$ and $j$, $R_{ij}^n$ is set equal to infinity. Equation 2.24 shows that $Y_{n,ij}$ is a symmetric $N \times N$ matrix.

Matrix equations suitable for determining GIC in a discretely-earthed system were independently derived by Lehtinen and Pirjola in 1985 [156]. Lehtinen-Pirjola method specifies the current to ground $[I^e]$ as the unknown (Eq. 2.25). The positive direction is assumed to be into the Earth.

$$[\mathbf{I}^e] = ([\mathbf{1}] + [\mathbf{Y}^n][\mathbf{Z}^e])^{-1}[\mathbf{J}^e], \tag{2.25}$$

where $[1]$ is the identity matrix; $[Y^n]$ is the admittance matrix specified as for nodal admittance matrix method; $[Z^e]$ is the earthing impedance matrix relating the currents to ground from node to the nodal voltages $v_k = \sum_{n=1}^{N} Z_{kn}^e i_n$; $[J^e]$ is the current vector source with elements given by summing the current sources directed into each node $J_k = \sum_{n=1}^{N} j_{nk}$.

Comparison of the nodal admittance matrix method and Lehtinen-Pirjola method shows that they are mathematically equivalent [106]. The modeling example is presented in [105].

In case two nodes are geographically close, so that the current to the ground from the first node produces a voltage drop that influences the other node. The Lehtinen-Pirjola method is a little more general compared to Nodal Admittance Matrix method by allowing the voltage of one node (earthing point) to be influenced by currents flowing through other nodes to/from the ground. The impact of the neighboring nodes is considered by off-diagonal elements in the impedance matrix. In practice, the off-diagonal elements do not play a major role [80]. Though an equivalent effect occurs when there are paths for the GIC flows through two or more transformer windings that share a common path through a substation grounding resistance, so that current through one transformer winding produced a voltage drop that influences the current flow through the other transformer winding [106]. As a measure to protect transmission lines from lightnings, the overhead shield wires are used. They perform as another path for the GIC flow. However, it is shown that their impact on the GIC distribution can be neglected [158].

Contrarily to power grids, pipelines are a continuously-earthed network. The applicability of the distributed-source transmission line theory for the GIC calculation in buried pipelines was shown in [159]. An important extension is the possibility of handling branches of the pipeline system was developed by [160]. The pipeline is treated as a transmission line containing a series impedance $Z$ determined by the resistance of the pipeline steel and a parallel admittance $Y$ associated with the conductivity of the coating; the geoelectric field affecting the pipeline forms the "distributed source" [120].

Much of the data for conductivity matrix $\mathbf{Y}$ is normally contained in power flow models or can be estimated. In contrast, the substation grounding resistance is usually not accurately available. The grounding resistance is the function of construction then it is a function of local soil parameters. This information is normally derived based on the sparse results of magnetotelluric surveys published in the geophysical papers. The paper [161] addresses the issue of the GIC sensitivity to the substation grounding resistance and provides an algorithm suitable for large power grid modeling.

Due to operational and construction requirements each voltage level has its own specific resistance. High-voltage lines have lower resistance, so they perform as the preferable GIC paths. Despite this fact, the GIC modeling by considering only high-voltage network gives inaccurate results. The first study to consider two voltage levels was done by [162]. The path for the GIC flows between voltage levels is through the galvanically connected power transformer windings. Therefore, as it is represented in [10] the analysis of the whole voltage levels spectrum is crucial in nodes with autotransformers as the nodes that share both high and low voltage level lines. The calculation error can be up to 20%, though ignoring the 110 kV voltage level does not have a very big effect [131]. Study of the extra-high voltage power grid in China (1,000 kV) showed similar results [164]. Autotransformer has no direct connection between high voltage winding and the neutral point. The series winding provides a connection between the high-voltage and low-voltage buses, and the shunt winding connects the low-voltage bus to neutral point.

Another type of power transformer that provides a path for the GIC is a two-winding power transformer which consists of two separate voltage windings connected to the neutral point. The neutral point is the junction between the two windings and the grounding. Three winding types exist: delta $\Delta$, star **wye** and star with a grounded neutral **wye$_0$**, the combination of which provide various transformer types. Both high-voltage and low-voltage windings of grid transformers are **wye$_0$**-connected due to safety reasons and designated as **wye$_0$-wye$_0$**. The low-voltage winding of the step-up transformer is designed as $\Delta$-winding and designated as $\Delta$-**wye**.

Δ-windings block GIC distribution, therefore are not going to be considered in the future analysis. The scheme of **wye₀**-Δ is shown in Fig. 2.23 together with the vector diagram.

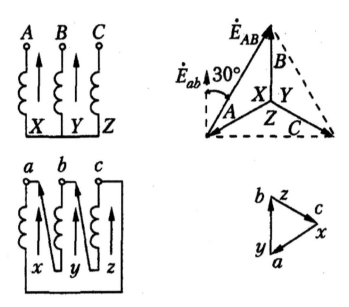

Figure 2.23: The scheme of **wye₀**-Δ connected power transformer on the right; its vector diagram on the left (Courtesy of Prof. Popov, V.)

One more approach to correlate the geomagnetic data with the GIC amplitudes is to produce GIC oriented forecasts. Two parameters were studied in [165], i.e. $A_p$ index, which is more linearly proportional to the size of the GMD, and geomagnetic field time derivative $dB/dt$. The method gives good accuracy up to ca. 89% for the nodes located in the mid and sub-aurora latitudes. The correlation for auroral zones decreases to ca. 50%. Such a difference can be due to the fact that Kp and correlated with it $A_p$ are derived from geomagnetic variations at sub-auroral and mid-latitude observatories. The other finding opened the discussion about how linear is the dependence between GIC and $A_p$ index, since the logarithmic power law showed better results. In general, the analysis using the magnetic field time derivative showed a better correlation especially for 3-hourly peak GIC values. It suggests that it is easier to predict the set of GIC variations rather than the detailed GIC variations themselves using 1 min values.

The reverse approach is proposed in [166] using the single value decomposition (SVD) technique. It was earlier shown that GICs are linearly dependent on the **E** field through the coefficient matrix **H**. Matrix **H** is determined only by network topology and resistances of the elements. The advantage of SVD technique is that it does not require any prior information about the network. The method estimates E-field through the GIC measurements without using any additional network information.

The SVD analysis also benefits from robustness to different types of uncertainties in the system.

According to [17], the GIC modeling from the known geoelectric field in the given conductor structure is the most developed link in the GMD risk assessment. The application readiness level for this link is 9, which means the sustained use of application system in decision-making context [168]. It should be noted that application readiness level 9 does not mean that no future work is required. The main present challenge is the accuracy of the initial data for the GIC calculation.

## 2.5 CONCLUSION

From the societal point of view, the rigorous GIC amplitude modeling is needed for the development of optimal mitigation procedures (Chapter 5). The level of technological system hardening is based on cost/risk ratio criteria. Overestimation of the accepted GIC amplitude range may bring sufficient economic expenses, whilst its underestimation may result in a huge economic loss. Apart from preparing to extremes, modeling is essential for situational space weather awareness which is performed on a continuous basis.

Physics-based modeling is the ultimate test for our understanding of any natural phenomenon. The whole space weather chain should be properly modeled starting from solar atmosphere down to the Earth's surface. It is a silver bullet for obtaining adequate range of GIC amplitudes for past, present and future superintense storms that especially pose a threat to the society well-being. It was shown that though the progress was made, there are still open questions in the physical principles of certain space weather manifestations. The most developed chain link is the GIC amplitude calculation in the given technological system located on the well-described territory. However, the skills of the chain links modeling should be pushed forward. It is important to note that the set of GIC amplitudes not only poses a challenge to geophysical community, but also to engineers, industry, and policy makers.

## REFERENCES

1. Eddington, A.S. (1988). *The Internal Constitution of the Stars*. Cambridge University Press, Cambridge.
2. Kamide, Y., Chian, A.C.-L. (2007). *Handbook of the Solar-Terrestrial Environment*. Springer Science & Business Media, Berlin, Heidelberg.
3. Apell, J.N., McNeill, K. (2019). Updated and validated solar irradiance reference spectra for estimating environmental photodegradation rates. *Environmental Science: Processes & Impacts*, 21:427–437.
4. Schematic drawing of the solar structure. Accessible on `https://hinode.nao.ac.jp/en/news/results/formation-mechanisms-of-the-solar-chromosphere-revealed-by-hinode-and-iris/` (accessed on 20.07.2020).
5. Bothmer, V., Zhukov, A. (2007). The Sun as the prime source of space weather. In *Space Weather: Physics and Effects*, 31–102. Springer, Berlin, Heidelberg.

6. Eddy, J.A. (1978). Historical evidence for the existence of the solar cycle. *The Solar Output and Its Variation*, 51–71. Provided by the SAO/NASA Astrophysics Data System.

7. Driel-Gesztelyi, L., Owens, M.J. (2020). Solar cycle. *Oxford Research Encyclopedia of Physics*. Oxford University Press, Oxford. DOI: 10.1093/acrefore/9780190871994.013.9.

8. Clette, F., Berghmans, D., Vanlommel, P., et al. (2007). From the Wolf number to the International Sunspot Index: 25 years of SIDC. *Advances in Space Research*, 40:919–928.

9. The sunspot number. Accessible on `https://spaceweather.com/glossary/sunspotnumber.html` (accessed on 20.07.2020).

10. Bothmer, V., Daglis, I.A. (2007). *Space Weather-Physics and Effects*. Springer Science & Business Media, Berlin Heidelberg.

11. Gibson, S. (2018). Solar prominences: theory and models. *Living Reviews in Solar Physics*, 15(1), 7.

12. Gibson, S.E., Fan, Y., Török, T., Kliem, B. (2006). The evolving sigmoid: Evidence for magnetic flux ropes in the corona before, during, and after CMES. *Space Science Reviews*, 124:131–144.

13. Gopalswamy, N. (2009). The CME link to geomagnetic storms. In *Proceedings of the International Astronomical Union*, 5:326–335. doi: 10.1017/S1743921309992870.

14. Tsuda, T., Fujii, R., Shibata, K., Geller, M.-A. (Eds.). (2009). *Climate and Weather of the Sun-Earth System (CAWSES): selected papers from the 2007 Kyoto symposium*. Terrapub.

15. Gopalswamy, N. (2018). Extreme solar eruptions and their space weather consequences. *Extreme Events in Geospace*, 37–63. Elsevier.

16. Gopalswamy, N. (2016). History and development of coronal mass ejections as a key player in solar terrestrial relationship. *Geoscience Letters*, 3(1), 1–18. doi: 10.1186/s40562-016-0039-2.

17. Kane, P.R. (2009). Early history of cosmic rays and solar wind–Some personal remembrances. *Advances in Space Research*, 44:1252–1255.

18. Bothmer, V. (2006). The solar atmosphere and space weather. *Solar System Update* 1–53. Springer, Berlin, Heidelberg.

19. Surkov, V., Hayakawa, M. (2014). *Ultra and Extremely Low Frequency Electromagnetic Fields*, 486. Tokyo: Springer Japan.

20. Ko, Y. K., Roberts, D. A., Lepri, S. T. (2018). Boundary of the slow solar wind. *The Astrophysical Journal*, 864(2), 139.

21. Zhao, L., Zurbuchen, T. H., Fisk, L. A. (2009). Global distribution of the solar wind during solar cycle 23: ACE observations. *Geophysical Research Letters*, 36(14).

22. Sanchez-Diaz, E., Rouillard, A.P., Lavraud, B., et al. (2016). The very slow solar wind: properties, origin and variability. *Journal of Geophysical Research: Space Physics*, 121:2830–2841.

23. Schwenn, R. (2001). Solar wind: Global properties. *Encyclopedia of Astronomy and Astrophysics*, 785998, 1–9.

24. Nishida, H., Ogawa, H., Inatani, Y. (2009). MHD Analysis of Force Acting on Dipole Magnetic Field in Magnetized Plasma Flow. *Computational Fluid Dynamics 2006*, 765–770. Springer, Berlin, Heidelberg.

25. Spreiter, J.R., Summers, A.L., Alksne, A.Y. (1966). Hydromagnetic flow around the magnetosphere. *Planetary and Space Science*, 14:223–253.

26. National Research Council (1993). *The National Geomagnetic Initiative*. The National Academies Press. Washington DC. doi: 10.17226/2238.
27. Russell, C.T., Strangeway, R.J., Zhao, C., et al. (2017). Structure, force balance, and topology of Earth's magnetopause. *Science*, 356:960–963.
28. Eastwood, J.P., Nakamura, R., Turc, L., et al. (2017). The scientific foundations of forecasting magnetospheric space weather. *Space Science Reviews*, doi: 10.1007/s11214-017-0399-8.
29. Freeman, M.P., Forsyth, C., Rae, I.J. (2019). The influence of substorms on extreme rates of change of the surface horizontal magnetic field in the United Kingdom. *Space Weather*, doi: 10.1029/2018SW002148.
30. Howard, R.A. (2006). *A Historical Perspective on Coronal Mass Ejections*. Geophysical Monograph-American Geophysical Union, 165.
31. Russel, C.T., McPherron, R.L. (1973). The magnetotail and substorms. *Space Science Reviews*, 15:205–266.
32. Tsurutani, B.T., Lakhina, G.S., Hajra, R. (2019). The physics of space weather/solar-terrestrial physics (STP): What we know now and what the current and future challenges are. *Nonlinear Processes in Geophysics Discussions*, doi: 10.5194/npg-2019-38.
33. Oler, C. (2004). Imperfect K index satisfactory for electric power industry. *Space Weather*, doi: 10.1029/2004SW000083.
34. Gopalswamy, N., Akiyama, S., Yashiro, S., et al. (2010). Coronal mass ejection from sunspot and non sunspot regions. *Magnetic Coupling Between the Interior and Atmosphere of the Sun*. Springer, Berlin.
35. Yermolaev, Y. I., Lodkina, I. G., Nikolaeva, N. S., Yermolaev, M. Y. (2018). Geoeffectiveness of Solar and Interplanetary Structures and Generation of Strong Geomagnetic Storms. *Extreme Events in Geospace*, 99–113. Elsevier.
36. Tsurutani, B.T. (2001). The interplanetary causes of magnetic storms, substorms and geomagnetic quiet. *Space Storms and Space Weather Hazards*, 103–130, Springer, Dordrecht.
37. Richardson, I.G., Cliver, E.W., Cane, H.V. (2001). Sources of geomagnetic storms for solar minimum and maximum conditions during 1972–2000. *Geophysical Research*, 28:2569–2572.
38. Tsurutani, B.T., Bruce, T., Gonzalez, W.D., et al. (2006). Corotating solar wind streams and recurrent geomagnetic activity: A review. *Journal of Geophysical Research: Space Physics, 111(A7)*. doi: 10.1029/2005JA011273.
39. Tsurutani, B.T., Gonzalez, W.D. (1997). The interplanetary causes of magnetic storms: A review. *GMS*, 98:77–89.
40. Richardson, I.G., Webb D.F., Zhang, J., et al. (2006). Major geomagnetic storms (Dst$\leq$ $-100$nT) generated by corotating interaction regions. *Journal of Geophysical Research*. doi: 10.1029/2005JA011476.
41. Weigel, R.S. (2010). Solar wind density influence on geomagnetic storm intensity. *Journal of Geophysical Research: Space Physics*. doi: 10.1029/2009JA015062.
42. Rostoker, G., Friedrich, E., Dobbs, M. (1997). Physics of magnetic storms. *Magnetic storms*, 98, 149–160. doi: 10.1029/GM098p0149.
43. Cliver, E.W., Feynman, J., Garrett, H.B. (1990). An estimate of the maximum speed of the solar wind, 1938–1989. *Journal of Geophysical Research: Space Physics*, 95:17103–17112.
44. Araki, T. (2014). Historically largest geomagnetic sudden commencement (SC) since 1868. *Earth, Planets and Space*, 66(1), 164. doi: 10.1186/s40623-014-0164-0.

45. Kamide, Y., Yokoyama, N., Gonzalez, W., et al. (1998). Two-step development of geomagnetic storms. *Journal of Geophysical Research*, 103:6917–6921.

46. Richardson, I.G., Zhang, J. (2008). History and development of coronal mass ejections as a key player in solar terrestrial relationship. *Geophysical Research Letters*. doi: 10.1029/2007GL032025 L06S07.

47. Taylor, J.R., Lester, M., Yeoman, T.K. (1994). A superposed epoch analysis of geomagnetic storms. *Annales Geophysicae*, 12:612–624. Springer-Verlag.

48. Tsurutani, B.T., Gonzalez, W.D., Tang, A.L.C., et al. (1995). The interplanetary causes of magnetic storms, substorms and geomagnetic quiet. *Journal of Geophysical Research*, 100: 21717–21733.

49. Silbergleit, V.M., Zossi de Artigas, M.M., Manzano, J.R. (2006). Austral electrojet indices derived for the great storm of March 1989. *Annals of Geophysics*. doi: 10.4401/ag-4046.

50. Huttunen, K.E., Koskinen, H.E.J., Schwenn, R. (2002). Variability of magnetospheric storms driven by different solar wind perturbations. *Journal of Geophysical Research: Space Physics*, 107(A7), SMP-20. doi: 10.1029/2001JA900171.

51. Barlow, W.H. (1849). On the spontaneous electrical currents observed in the wires of the electric telegraph. *Philosophical Transactions of the Royal Society of London*, 139:61–72.

52. New York Times (1921) Sunspot credited with rail tie-up. *New York Times*, May 16, 1921, 2.

53. Davidson, W.F. (1940). The magnetic storm of March 24, 1940. Effects in the power system. *Edison Electric Institute Bulletin*, 8:365–366.

54. Harang, L. (1951). The aurorae. *Quarterly Journal of the Royal Metrological Society* 78.

55. Acres Consulting Services Ltd. (1975). *Study of the disruption of electric power systems by magnetic storms*. Earth Phys. Branch Open File 77-19, Department of Energy, Mines and Resources, Ottawa.

56. Slothower, J.C., Albertson, V.D. (1967). The effects of solar magnetic activity on electric power systems. *Journal of Minn. Academy of Science*, 34:94–100.

57. Boteler, D.H. (2019). A twenty-first century view of the march 1989 magnetic storm. *Space Weather*. doi: 10.1029/2019SW002278.

58. Bolduc, L. (2002). GIC observations and studies in the Hydro-Québec power system. *Journal of Atmospheric and Solar-Terrestrial Physics*, 64:1793–1802.

59. Elovaara, J., Lindblad, P., Viljanen, A., et al. (1992). Geomagnetically induced currents in the Nordic power system and their effects on equipment, control, protection and operation. Paper 36-301. In *Proceedings of the CIGRE 1992 Session*, 30 August–5 September, Paris.

60. Pulkkinen, A., Lindahl, S., Viljanen, A., Pirjola, R. (2005). Geomagnetic storm of 29-31 October 2003: Geomagnetically induced currents and their relation to problems in the Swedish high-voltage power transmission system. *Space Weather*. doi: 10.1029/2004SW000123.

61. Belov, A.V., Gaidash, S.P., Eroshenko, E.A., et al. (2007). Effects of strong geomagnetic storms on Northern railways in Russia. In *Proceedings of 7th International Symposium and Exhibition on Electromagnetic Compatibility and Electromagnetic Ecology (IEEE EMC)*, Sankt-Petersburg. doi: 10.1109/EMCECO.2007.4371710.

62. Wik, M., Pirjola, R., Lundstedt, H., et al. (2009). Space weather events in July 1982 and October 2003 and the effects of geomagnetically induced currents on Swedish technical

systems. *Annales Geophysicae*, 27:1775–1787.
63. Thaduri, A., Galar, D., Kumar, U. (2020). Space weather climate impacts on railway infrastructure. *International Journal of System Assurance Engineering and Management*, 11(2), 267–281.
64. Erinmez, I.A., Kappenman, J.G., Radasky, W.A. (2005). Management of the geomagnetically induced current risks on the national grid company's electric power transmission system. *Journal of Atmospheric and Solar-Terrestrial Physics*. doi: 10.1016/S1364-6826(02)00036-6.
65. Liu, C.M., Liu, L.G., Pirjola, R. (2009). Geomagnetically induced currents in the high-voltage power grid in China. *IEEE Transactions on Power Delivery*, 24(4), 2368–2374. doi: 10.1109/TPWRD.2009.2028490.
66. Gaunt, C.T., Coetzee, G. (2007). Transformer failures in regions incorrectly considered to have low GIC-risk. *2007 IEEE Lausanne Power Tech*, 807–812.
67. Gaunt, C.T. (2014). Reducing uncertainty responses for electricity utilities to severe solar storms. *Journal of Space Weather and Space Climate*, 4, A01. doi: 10.1051/swsc/2013058.
68. Trichtchenko, L., Zhukov, A., Van der Linden, L., et al. (2007). November 2004 space weather events: Real time observations and forecasts. *Space Weather*, 5(6), doi: 10.1029/2006SW000281.
69. Thomson, A., McKay, A.J., Clarke, E., et al. (2005). Surface electric fields and geomagnetically induced currents in the Scottish Power grid during the 30 October 2003 geomagnetic storm. *Space Weather*, 3(11), doi: 10.1029/2005SW000156.
70. MacMillan, R. K., Keefe, R. L., Perala, R. A. (2000). Geomagnetic Storms and the Possible Effects on the Railroad System. In *AREMA Proceedings of the 2000 Annual ConferenceAmerican Railway Engineering and Maintenance-of-Way Association*.
71. Brekke, P. (2004). Space weather effects. In *Proceedings of First European Space Weather Week (ESWW), SA-ESTEC, Noordwijk*.
72. Eroshenko, E.A., Belov, A.V., Boteler, D., et al. (2010). Effects of strong geomagnetic storms on Northern railways in Russia. *Advances in Space Research*, 46(9), 1102–1110. doi: 10.1016/j.asr.2010.05.017.
73. Ptitsyna, N.G., Tyasto, M.I., Kassinskii, V.V., et al. (2007). Do Natural Magnetic Fields Disturb Railway Telemetry? In *Proceedings of 7th International Symposium on IEEE Electromagnetic Compatibility and Electromagnetic Ecology*, 288–290. IEEE.
74. Sakharov, Ya.A., et al. (2010). Effects of geomagnetic disturbances on Oktyaborskaya railway in Russia. In *Proceedings of 7th European Space Weather Week (ESWW), SA-ESTEC, Brugge*.
75. Papavinasam, S. (2013). *Corrosion Control in the Oil and Gas Industry*. Elsevier, USA.
76. Boteler, D.H., Trichtchenko, L. (2000). *International study of telluric currents on pipelines, Final Report*. Geological Survey of Canada, Open File 3050, Canada.
77. Fernberg, P.A., Samson, C., Boteler, D.H., et al. (2007). Earth conductivity structures and their effects on geomagnetic induction in pipelines. *Annales Geophysicae*. doi: 10.5194/angeo-25-207-2007.
78. Osella, A., Favetto, A. (2000). Effects of soil resistivity on currents induced on pipelines. *Journal of Applied Geophysics*, 44:303–312.
79. Boteler, D.H. (2000). Geomagnetic effects on the pipe-to-soil potentials of a continental pipeline. *Advances in Space Research*, 26(1). doi: 10.1029/2009SW000553.
80. Pirjola, R., Viljanen, A., Amm, O., Pulkkinen, A. (1999). Power and pipelines (ground systems). In *Proceedings of a Workshop on Space Weather, November 1998, ESA WPP*.

81. Hopper, A.T., Thompson, R.E. (1970). *Earth Current Effects on Buried Pipelines: Analysis of Observations of Telluric Gradients and Their Effects*. American Gas Association.

82. Lundstedt, H. (1993). Solar caused potential in gas-pipelines in southern Sweden. *Solar-Terrestrial Predictions–IV 1*,233.

83. Lundstedt, H. (2006). The sun, space weather and GIC effects in Sweden. *Advances in Space Research*, 37(6), 1182–1191. doi: 10.1016/j.asr.2005.10.023.

84. Hejda, P., Bochnicek, J. (2005). Geomagnetically induced pipe-to-soil voltages in the Czech oil pipelines during October-November 2003. *Annales Geophysicae*, 23(9), 3089–3093. doi: 10.5194/angeo-23-3089-2005.

85. Pulkkinen, A., Viljanen, A., Pajunpää, K. et al. (2001). Recordings and occurrence of geomagnetically induced currents in the Finnish natural gas pipeline network. *Journal of Applied Geophysics*, 48(4), 219–231. doi: 10.1016/S0926-9851(01)00108-2.

86. Jankowski, J., Pirjola, R., Ernst, T. (1986). Homogeneity of magnetic variations around the Nurmijärvi observatory. *Geophysica*, 22:3–13.

87. Viljanen, A., Koistinen, A., Pajunpää, K., Pirjola, R., Posio, P., Pulkkinen, A. (2010). Recordings of geomagnetically induced currents in the Finnish natural gas pipeline–summary of an 11-year period. *Geophysica*, 46:59–67.

88. Viljanen, A., Pulkkinen, A., Pirjola, R., Pajunpää, K., Posio, P., Koistinen, A. (2006). Recordings of geomagnetically induced currents and a nowcasting service of the Finnish natural gas pipeline system. *Space Weather*, 4: S10004.

89. Ingham, M., Rodger, C. (2018). Telluric field variations as drivers of variations in cathodic protection potential on a natural gas pipeline in New Zealand. *Space Weather*, 16:1396–1409.

90. Ingham, M. (1993). Analysis of variations in cathodic protection potential and corrosion risk on the natural gas pipeline at Dannevirke. *Report prepared for Natural Gas Corporation*.

91. Marshall, R.A., Waters, C.L., Sciffer, M.D. (2010). Spectral analysis of pipe-to-soil potentials with variations of the Earth's magnetic field in the Australian region. *Space Weather*, 8:1–13.

92. Siscoe, G., Crooker, N.U., Clauer, C.R. (2006). Dst of the Carrington storm of 1859. *Advances in Space Research*, 38: 173–179.

93. Steinhilber, F., Abreu, J., Beer, J., et al. (2012). 9,400 years of cosmic radiation and solar activity from ice cores and tree rings. In *Proceedings of the National Academy of Sciences*, 109: 5967–5971.

94. Wolff, E.W., Jones, A.E., Bauguitte, S.J.B., Rhian, A. (2008). The interpretation of spikes and trends in concentration of nitrate in polar ice cores, based on evidence from snow and atmospheric measurements. *Atmospheric Chemistry and Physics*, 8(18), doi: 10.5194/acp-8-5627-2008.

95. Baker, D.N., et al. (2013). A major solar eruptive event in July 2012: Defining extreme space weather scenarios. *Space Weather*, 11:585–591.

96. Pulkkinen, A., Viljanen, A. (2007). The complex spatiotemporal dynamics of ionospheric currents. *COST 724 final report*.

97. Zheng, K., Trichtchenko, L., Pirjola, R., et al. (2013). Effects of geophysical parameters on GIC illustrated by benchmark network modeling. *IEEE Transactions on Power Delivery*, 28:1183–1191.

98. Thomson, A., McKay, A.J., Viljanen, A. (2009). A review of progress in modelling of induced geoelectric and geomagnetic fields with special regard to induced currents. *Acta Geophysica*, 57:209–219.

99. Carter, B.A., Yizengaw, E., Pradipta, R., et al. (2016). Geomagnetically induced currents around the world during the 17 March 2015 storm. *Journal of Geophysical Research: Space Physics*, 121:10496–10507.
100. Rasson, J.L., Toh, H., Yang, D. (2011). The global geomagnetic observatory network. *Geomagnetic Observations and Models*, 1–25.
101. Love, J.J., Chulliat, A. (2013). An international network of magnetic observatories. *Eos, Transactions American Geophysical Union*, 94:373–374.
102. Jankowski, J., Sucksdorff, C. (1996). Guide for magnetic measurements and observatory practice. *International Association of Geomagnetism and Aeronomy Warsaw*.
103. Kerridge, D. (2001). INTERMAGNET: Worldwide near-real-time geomagnetic observatory data. In *Proceedings of the workshop on space weather, ESTEC, 34*.
104. Love, J.J. (2008). Magnetic monitoring of Earth and space. *Physics Today*, 61:31–37.
105. Boteler, D., Pirjola, R. (2017). Modeling geomagnetically induced currents. *Space Weather*, 15:258–276.
106. Boteler, D., Pirjola, R. (2014). Comparison of methods for modelling geomagnetically induced currents. *Annales Geophysicae*, 32:1177–1187.
107. Viljanen, A., Pulkkinen, A., Amm, O., et al. (2004). Fast computation of the geoelectric field using the method of elementary current systems and planar Earth models. *Annales Geophysicae*, 22:101–113.
108. Haines, G.V., Torta, J.M. (1994). Determination of cquivalent current sources from spherical cap harmonic models of geomagnetic field variations. *Geophysical Journal International*, 118:499–514.
109. Viljanen, A., Nevanlinna, H., Pajunpaa, K., et al. (2001). Time derivative of the horizontal geomagnetic field as an activity indicator. *Annales Geophysicae*, 19:1107–1118.
110. Pulkkinen, A., Rastatter, L., Kuznetsova, M., et al. (2013). Community-wide validation of geospace model ground magnetic field perturbation predictions to support model transition to operations. *Space Weather*, 11:369–385.
111. Welling, D.T., Ngwira, C.M., Opgeoorth, H. (2018). Recommendations for next-generation ground magnetic perturbation validation. *Space Weather*, 16: 1912–1920.
112. Pulkkinen, A., Rastaetter, L., Kuznetsova, M. (2010). Systematic evaluation of ground and geostationary magnetic field predictions generated by global magnetohydrodynamic models. *Journal of Geophysical Research*, 115(A3). doi: 10.1029/2009JA014537.
113. Pulkkinen, A., Viljanen, A., Pirjola, R. (2006). Estimation of geomagnetically induced current levels from different input data. *Space Weather*, 4(8). doi: 10.1029/2006SW000229.
114. Toth, G., Meng, X., Gombosi, I., et al. (2014). Predicting the time derivative of local magnetic perturbation. *Geophysical Research Space Physics*, 119:310–321.
115. Jolliffe, I.T., Stephenson, D.B. (2012). *Forecast Verification: A Practitioner's Guide in Atmospheric Science*. John Wiley & Sons, Chichester.
116. Schrijver, C., Carolus, J., Kauristie, K., et al. (2015). Understanding space weather to shield society: A global road map for 2015–2025 commissioned by COSPAR and ILWS. *Advances in Space Research*, 55:2745–2807.
117. Seth, J., McCarron, E. (2016). White house releases national space weather strategy and action plan. *Space Weather*, 14:54–55.
118. Pirjola, R. (2000). Space weather effects on technological systems on the ground. In *Proceedings. Asia-Pacific Conference on Environmental Electromagnetics. CEEM'2000*, 217–221.

119. Pirjola, R. (1982). Electromagnetic induction in the earth by a plane wave or by fields of line currents harmonic in time and space. *Geophysica*. Geophysical Society of Finland, Finland.

120. Pirjola, R. (2002). Review on the calculation of surface electric and magnetic fields and of geomagnetically induced currents in ground-based technological systems. *Surveys in geophysics*, 23:71–90.

121. Cagniard, L. (1953). Basic theory of the magneto-telluric method of geophysical prospecting. *Geophysics*, 18:605–635.

122. Pirjola, R. (1985). On currents induced in power transmission systems during geomagnetic variations. *IEEE Transactions on Power Apparatus and Systems*, 10:2825–2831.

123. Viljanen, A., Pirjola, R. (1989). Statistics on geomagnetically-induced currents in the Finnish 400kV power system based on recordings of geomagnetic variations. *Journal of Geomagnetism and Geoelectricity*, 41:411–420.

124. Pirjola, R., Viljanen, A. (1989). On geomagnetically-induced currents in the Finnish 400 kV power system by an auroral electrojet current. *IEEE Transactions on Power Delivery*, 4:1239–1245.

125. Towle, J.N., Prabhakara, F.S., Ponder, J.Z. (1992). Geomagnetic effects modelling for the PJM interconnection system. Part I: Earth surface potential computation. *IEEE Transactions on Power Systems*, 7:949–955.

126. Pirjola, R., Viljanen, A., Boteler, D. (1999). Series expansions for the electric and magnetic fields produced by a line or sheet current source above a layered Earth. *Radio Science*, 34:269–280.

127. Pirjola, R. (1998). Modelling the electric and magnetic fields at the Earth's surface due to an auroral electrojet. *Journal of Atmospheric and Solar-Terrestrial Physics*, 60:1139–1148.

128. Thomson, D.J., Weaver, J.T. (1975). The complex image method approximation for induction in a multilayered Earth. *Journal of Geophysical Research*, 80: 123–129.

129. Amm, O. (1995). Direct determination of the local ionospheric Hall conductance distribution from two-dimensional electric and magnetic field data: Application of the method using models of typical ionospheric electrodynamic situations. *Journal of Geophysical Research*, 100:21473–21488.

130. Beggan, C.D. (2015). Sensitivity of geomagnetically induced currents to varying auroral electrojet and conductivity models. *Earth, Planets and Space*, 67(1). doi: 10.1186/s40623-014-0168-9.

131. Horton, R., Boteler, D., Pirjola, R., et al. (2012). A test case for the calculation of geomagnetically induced currents. *IEEE Transactions on Power Delivery*, 27(4). doi: 10.1109/TPRWD.2012.2206407.

132. Pirjola, R. (2013). Practical model applicable to investigating the coast effect on the geoelectric field in connection with studies of geomagnetically induced currents. *Advances in Applied Physics*, 1:9–28.

133. Marti, L., Yiu, C., Rezaei-Zare, A., et al. (2014). Simulation of geomagnetically induced currents with piecewise layered-earth models. *IEEE Transactions on Power Delivery*, 29:1886–1893.

134. Dong, B., Wang, Z., Boteler, D., et al. (2013). Review of earth conductivity structure modelling for calculating geo-electric fields. *IEEE Power & Energy Society General Meetings*, 1–5.

135. Wait, J.R. (1953). Propagation of radio waves over a stratified ground. *Geophysics*, 18:416–422.

136. Price, A.T. (1967). Electromagnetic induction within the earth. International *Geophysics*, 11:235–298.
137. Parkinson, W.D., Jones, F.W. (1979). The geomagnetic coast effect. Reviews of *Geophysics*, 17:1999–2015.
138. Wait, J.R., Spies, K.P. (1974). Magneto-telluric fields for a segmented overburden. *Journal of Geomagnetism and Geoelectricity*, 26:449–458.
139. Ferguson, I. (2012) Instrumentation and field procedures. *The Magnetotelluric Method: Theory and Practice*, 421.
140. Chave, A.D. (2012). Estimation of the magnetotelluric response function. *The Magnetotelluric Method: Theory and Practice*, 165–218.
141. Rosi, W.L., Mackie, R.L. (2012). The inverse problem. *The Magnetotelluric Method: Theory and Practice*, 347–414.
142. Ferguson, I.J., Jones, A.G., Chave, A.D. (2012). Case histories and geological applications. *The Magnetotelluric Method: Theory and Practice*, 480–544.
143. Bedrosian, P.A., Love, J.J. (2015). Mapping geoelectric fields during magnetic storms: Synthetic analysis of empirical United States impedances. *Geophysics*, 42:10160–10170.
144. Chave, A.D., Jones, A.G. (2012). *The Magnetotelluric Method: Theory and Practice*. Cambridge University Press, Cambridge.
145. Siripunvaraporn, W. (2012). Three-dimensional magnetotelluric inversion: an introductory guide for developers and users. *Surveys in geophysics*, 33:5–27.
146. Boteler, D.H. (2014). The evolution of Quebec Earth models used to model geomagnetically induced currents. *IEEE Transactions on Power Delivery*, 30:2171–2178.
147. Boteler, D.H. (1994). Geomagnetically induced currents: Present knowledge and future research. *IEEE Transactions on Power Delivery*, 9:50–58.
148. Ferguson, I. J., Odwar, H. D. (1997). Review of conductivity soundings in Canada (Appendix 3). *Geomagnetically Induced Currents: Geomagnetic Hazard Assessment, phase II, 357 T 848A*, 3, 1–121.
149. Kuvshinov, A., Olsen, N. (2006). A global model of mantle conductivity derived from 5 years of CHAMP, Ørsted, and SAC-C magnetic data. *Geophysical Research Letters*, 33(18), doi: 10.1029/2006GL027083.
150. Olsen, N., Kuvshinov, A. (2004). Modeling the ocean effect of geomagnetic storms. *Earth, planets and space, 56(5)*. doi: 10.1186/BF03352512.
151. Ngwira, C.M., Pulkkinen, A.A., Bernabeu, E., et al. (2014). Characteristics of extreme geoelectric fields and their possible causes: Localized peak enhancements. *Geophysical Research Letters*, 42:6916–6921.
152. Ramleth, K.J. (1982). Geomagnetiske forstyrrelser. Innvirkning pâ Televerkets linjenett. *Telektronikk*, 10–12.
153. Zheng, K., Liu, L., Boteler, D.H., et al. (2013). Calculation analysis of geomagnetically induced currents with different network topologies. *IEEE Power & Energy Society General Meeting*, 1–4. doi: 10.1109/PESMG.2013.6672372.
154. Kappenman, J. (2010). Low-frequency protection concepts for the electric power grid: geomagnetically induced current (GIC) and E3 HEMP mitigation. *FERC, Metatech Corporation*.
155. Boteler, D.H., Pirjola, R., Blais, C., et al. (2014). Development of a GIC Simulator. *IEEE PES General Meeting— Conference & Exposition*, 1–5. doi: 10.1109/PESGM.2014.6939778.

156. Lehtinen, M., Pirjola, R. (1985). Currents produced in earthed conductor networks by geomagnetically-induced electric fields. *Annales Geophysicae*, 3:479–484.

157. Pirjola, R. (2008). Effects of interactions between stations on the calculation of geomagnetically induced currents in an electric power transmission system. *Earth Planets Space*, 60:743–751.

158. Pirjola, R. (2007). Calculation of geomagnetically induced currents (GIC) in a high-voltage electric power transmission system and estimation of effects of overhead shield wires on GIC modelling. *Journal of Atmospheric and Solar-Terrestrial Physics*, 69:1305–1311.

159. Boteler, D.H. (1997). Distributed-source transmission line theory for electromagnetic induction studies. In *Proceedings of the 1997 Zurich EMC Symposium*, 401–408.

160. Pulkkinen, A., Pirjola, R., Boteler, D., et al. (2001). Modelling of space weather effects on pipelines. *Journal of Applied Geophysics*, 48:233–256.

161. Bui, U., Overbye, T., Shetye, K. (2013). Geomagnetically induced current sensitivity to assumed substation grounding resistance. *2013 North American Power Symposium (NAPS)*, 1–6.

162. Makinen, T. (1993). *Geomagnetically Induced Currents in the Finish Power Transmission System*. Finnish Meteorological Institute, Helsinki.

163. Torta, J., Marsal, S., Quintana, M. (2014). Assessing the hazard from geomagnetically induced currents to the entire high-voltage power network in Spain. *Earth, Planets and Space*, 66(1). doi: 10.1186/1880-5981-66-87.

164. Guo, S., Liu, L., Pirjola, R., et al. (2015). Impact of the EHV power system on geomagnetically induced currents in the UHV power system. *IEEE Transactions on Power Delivery*, 30(5), doi: 10.1109/TPRWD.2014.2381248.

165. Trichtchenko, L., Boteler, D.H. (2004). Modeling geomagnetically induced currents using geomagnetic indices and data. *IEEE Transactions on Plasma Science*, 32(4). doi: 10.1109/TPS.2004.830993.

166. Kazerooni, M., Zhu, H., Overbye, T.J. (2015). Singular value decomposition in geomagnetically induced current validation. *IEEE Power & Energy Society: Innovative Smart Grid Technologies*, 1–5. doi: 10.1109/ISGT.2015.7131889.

167. Pulkkinen, A., Bernabeu, E., Thomson, A., et al. (2017). Geomagnetically induced currents: Science, engineering, and applications readiness. *Space Weather*, 15:828–856.

168. Nelson, E.J., Pulla, S.T., Matin, M.A., et al. (2019). Enabling Stakeholder Decision-Making With Earth Observation and Modeling Data Using Tethys Platform. *Frontiers in Environmental Science*. doi: 10.3389/fenvs.2019.00148.

169. Boteler, D.H., Lackey, A.J.C., Marti, L., et al. (2013). Equivalent circuits for modelling geomagnetically induced currents from a neighbouring network. *IEEE Power & Energy Society General Meeting*, 1–5.

# 3 Geomagnetically Induced Currents Observation

## CHAPTER CONTENTS

3.1 Methods and equipment for the geomagnetically induced current
observation.................................................................................................... 65
3.2 Worldwide observation of geomagnetically induced currents........................ 66
3.3 Geomagnetically induced currents observation in north-west of Russia ........ 69
3.4 Conclusion.................................................................................................... 74

EXPLOSIVE events on the Sun may result in the GIC flows over conductive technological systems on the Earth's surface. GICs are the solar storm's echoes on Earth, which can result in huge economic loss. The GIC observation at various sites across the globe helps to validate the GIC calculation models, to better correlate the ionospheric-magnetospheric conditions (see Chapter 2) with the generated GICs, and to estimate extreme GIC values whose proper evaluation is crucial for the mitigation procedure planning (see Chapter 5). This chapter discusses the techniques used for the GIC observation in different countries, describes the historical background and achieved results, gives examples of extreme registered GICs, and an overview of the physical causes. In addition, the chapter explores the GIC observation in north-west part of Russia, which is not well known compared to other countries.

## 3.1 METHODS AND EQUIPMENT FOR THE GEOMAGNETICALLY INDUCED CURRENT OBSERVATION

The GIC measurement equipment is normally installed in the power transformers grounding cables, since no current flows over the grounding cable in the normal balanced state. Under unbalanced conditions, both GIC and AC present in the grounding wire, but AC can be easily filtered. The ultra-low frequency of the GIC ($10^{-4}$–$10^{-1}$ Hz) significantly differs from the nominative power grid frequency (50/60 Hz) or ultra-high frequency typical for transient processes. Predominantly, GICs are measured in the neutral-to-ground connection of the power transformer using instruments such as Hall-effect sensors, shunt resistors, or clamp ammeters. Hall-effect is named after the American physicist, Edwin H. Hall, who in 1879 proposed that a magnetic field applied to a current-carrying conductor would generate a state of stress in the conductor in the way that the electricity is pressed toward one end of the wire. The concept of using the phasor measurement unit (PMU) for the GIC monitoring was proposed in [1].

The design of low-cost GIC monitor system was proposed in [2]. The specifications were as following: the GIC monitoring system should be able to measure DC values up to 150 A, beyond which saturation level is acceptable; the sampling frequency is flexible but not less than 1 reading per 5 seconds; it should be clamp on with a diameter of at least 70 mm; it needs to perform a continuous sampling and provide sufficient data storage; it has to be robust to extreme environmental temperatures. The proposed system block diagram consists of the following blocks: temperature and Hall-effect sensors connected through band pass filter to microcontroller as inputs, microcontroller block is connected to energy supply block represented by 12V DC supply source. These blocks are universal, and typical for other monitoring systems.

The differential magnetometer method (DMM) targets the GIC estimation in a power line. The GIC is evaluated from the difference between two magnetic field measurements: directly underneath the power line, and at a distance from the power line, where the magnetic field of the GIC induced in the transmission line has a negligible effect. The DMM was first applied for the GIC measurement in the pipelines (Alaska gas pipeline) [3]. The same approach was used for the GIC measurement in the Finnish gas pipeline [4]. The DMM was used for the GIC estimation in transmission lines in South Africa. The field test results proved the ability of the DMM in estimating the GIC, and the obtained data can also serve as a backup when the transformer GIC measurements are not available due to technical faults [5], [6].

Another approach is to model GICs instead of measuring them by following the method described in Section 2.4. Such activities are performed in both hemispheres in various countries regardless to their latitudes. The list includes, but is not limited to, Uruguay [7], Australia [8], Kazakhstan [9], Spain [10], Turkey [11]. The obtained GIC amplitudes have limiting accuracy due to the model imperfection as it was shown in [12]. The GIC modeling principles and aspects are discussed in Chapter 2.

## 3.2  WORLDWIDE OBSERVATION OF GEOMAGNETICALLY INDUCED CURRENTS

In 1977 a collaborative research between the Finnish Meteorological Institute (FMI) and the Imatran Voima Oy (later Fingrid Oyj) power company on the GIC impact on Finnish power grid was launched. The data recordings already cover nearly four 11-year solar cycles. The research of GICs in the Finnish power system includes theoretical modeling of the geoelectric field, calculations and measurements of the GIC, and the statistical analyses of the GIC occurrence [13]. Only 400 kV network was in focus of this study, hence the studied scheme was extended to 220 kV grid as well. The modern 400 kV grid does not differ from the one constructed in the late 1970s (Fig. 3.1). The recordings of GICs in the earthing wire of the 400 kV Huutokoski node started in 1977 as 10-s mean values [14].

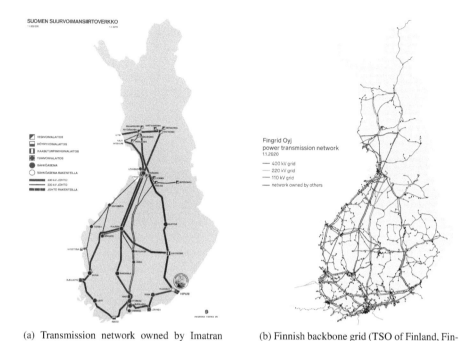

(a) Transmission network owned by Imatran Voima (later part of Fingrid Oyj) in the late 1970s

(b) Finnish backbone grid (TSO of Finland, Fingrid Oyj) as of early 2020s

Figure 3.1: Finnish high voltage power grid models [Credit: Power transmission network of Fingrid Oyj, TSO of Finland] (Courtesy of Fingrid Oyj)

The GIC amplitudes measured at the beginning of the recordings and the nowadays data exhibit clear differences in values. It could be due to the power grid characteristics change. For instance, the largest so far registered GIC is 201 A, i.e. 67 A per phase (as a 1-min mean value) in the earthing lead of the 400 kV transformer neutral at 400 kV Rauma substation on March 24, 1991 [15]. Meantime, the data obtained during the Halloween event shows that GIC of only ~40 A was measured in the same node. It should also be noted that the features of the October/November 2003 and March 1991 geomagnetic storms were quite different [16].

Efforts to identify the correlation between the GICs and the magnetic indices were done. [17] showed that at Boulder (magnetic latitude about 49° N) the range of the time derivative of the magnetic field plotted versus the K index indicates an ambiguity. Similar analysis was also done in Finland on the GICs recorded at Mäntsälä node versus magnetic field observations at Nurmijärvi observatory. In majority of events, the increase in the maximum GIC correlates with the magnetic activity rise, except the sudden storm commincement (SSC) events with relatively small amplitudes, but rapid magnetic field variations producing high GICs [18].

The problem of GMD impact on power grid operation goes beyond the territory of a single country. A special attention has been paid to collaborative GIC observations on a continent-scale [19]. GICs are recorded in other European countries such as Sweden [20] and Spain [21] as well. Since Chinese power grid was also found to be vulnerable to negative GIC effects, the GIC observation was launched there too [22, 23]. The GIC observation in extra-high voltage 765 kV transmission network in Korea started in 2002 [24].

The Hydro-Québec event boosted the development of GIC measurement network in North America. The Electric Power Research Institute has established a monitoring network called SUNBURST to collect high quality, readily accessible data related to GIC associated with GMDs [25]. The initial map is represented in [25]. Currently, the monitoring system includes approximately 50 strategically positioned nodes that are collecting GIC data. Thirty-seven nodes existed back in 2007. SUNBURST gathers data such as transformer phase currents, voltages, neutral currents, and hotspot temperatures, as well as electric and magnetic fields, which can be used by utilities to monitor online their own GIC values in real time and compare Earth currents for many other sites to gain a perspective concerning the magnitude and proximity of any unfolding solar storm [26]. The recorded data is available over a period as early as 1999. The Storm Catalog gives the project members a quick access to comprehensive information about significant events. The collection of maximum neutral currents over the last two solar cycles is given in [26]. SUNBURST generates insights into how GICs develop during a solar storm and how this data relates to prior observations.

GIC observation activities also take place in the Southern Hemisphere. The GIC monitoring in South Africa was launched in 2001 at the Grassridge electrical substation (33.7° S, 25.6° E) under the EPRI Sunburst project [27]. The Grassridge 400/220/132 kV substation, on which a single unit of 400/132 kV, 500 MVA transformer is installed and monitored, is located on the South-East coast of South Africa. Later, measurements at the Hydra substation were also started. It is a 400/220/132/22 kV substation located in the central part of South Africa. The monitoring equipment is installed on a three-phase, five-limb 400/132/15 kV, 240 MVA autotransformer [28]. This substation has experienced reactor and transformer failure in the past which could be related to GICs [29].

Two high voltages transmission lines in Brazil were chosen for the observation: Itumbiara São-Simão and Pimenta-Barreiro. Both of them have a length more than 150 km and do not contain series compensation devices. Besides, they are located in the east-west direction. The systematic measurements started in the early 2000s at Itumbiara substation. The Hall detectors for the current meter were installed directly on the surface of the cable linking the neutral point of the 500 kV transformer and the Earth [30]. The obtained GIC amplitudes for the November 2004 GMD were found to be around 15–20 A.

The archive of measured GICs amplitudes is a valuable basis for data verification used for the GIC modeling. The analysis performed in Austria showed a good correlation (linear relation coefficient $r > 0.8$) of data sets: measured and registered. The transformer neutral point measurements are conducted at four substations, geographically distributed over Austria. Nevertheless, it is important to note that there was no severe event since 2016, since the measurements started [31]. Alternative parameter (P) for evaluating how skillful is the model was introduced in [32]. The fact that P is a simple score has the advantage of providing a practical and effective idea of the model performance, but this fact limits its benefits at the same time. Its correlation with r is also given. Another task is the usage of registered data for training extreme-value models. More information on existing extreme-value models and forecasted values is given in Section 6.3. The verification of GICs extreme values was carried out for Hokkaido, Japan [33]. The estimated maximum values of the GICs were more than half lower than the observed ones [34].

The registration results may be used for scientific purposes. This may facilitate new discoveries. It is proposed to establish a repository in close collaboration with the industry for collecting data from a number of stations with wide geographical coverage for special geomagnetic storm events of interest [35]. For the optimal data processing, application of the uniform registration standards is required. The modern recommendations are as following:

1. the GIC recording with a 1 s or higher temporal cadence;
2. the measurement equipment should have a sensitivity of 0.1 A or higher;
3. include polarity of the current in the recordings;
4. carry out recordings with GPS synchronized time-steps;
5. the records should be stored continuously with GPS synchronized time-steps.

The quality and completeness of the data should be checked, validated, and approved. The lessons learned from incomplete and inaccurate data collection of US bulk power grid failures should be adopted whilst the GIC repository creation. The historical changes in the requirements and list of event characteristics limit the accuracy of further analysis and the event post-modeling [36]. NERC began collecting reliability data in 1984 and made it mandatory in 2007. Another database is the US Department of Energy initiative through the EIA-417 form established in the late 1970s.

## 3.3 GEOMAGNETICALLY INDUCED CURRENTS OBSERVATION IN NORTH-WEST OF RUSSIA

Compared to other countries located in the aurora region, the GIC measurement in North-West of Russia was started later, though the problem of space weather impact on power grids was in the research scope since 1980s. The first project was launched by the JSC "Energosetproekt", Moscow, together with research institute of Physical and Technical Problems of the North, Apatity, in 1986. However, the project lasted only for two years. In 1994, a corresponding member of the Russian Academy of Sciences, Prof. M.V. Kostenko, initiated an informal research group to study the impact of GMD on the power plants located in North-West of Russia. The co-members of the research group included representatives of the Centre for Physical and Technical Problems of Energy of the North, Polar Geophysical Institute of the Kola Scientific Centre of the Russian Academy of Science, Arctic and Antarctic scientific institute, and 26 Central Research institutes of the Ministry of Defense of Russian Federation. In the course of ten years, the group members developed the scientific and methodological foundation for assessing the GMD impact and ensuring the protection of power grids at the moment of GMD appearance.

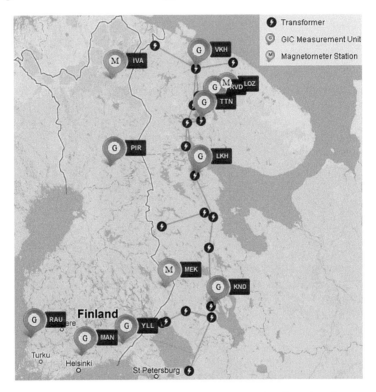

Figure 3.2: The GIC measurement network in the North-West Russia (Courtesy of Selivanov, V.)

The start of a systematic research on the GIC impact on power grid in North-West part of Russia was in 2003 [37]. The continuous GIC measurement on the set of high voltage substations started in 2011 within the EURISGIC project [38]. The measurement equipment is currently installed on the five substations along the 330 kV transmission line (Fig. 3.2). Thanks to its location, it gives a unique opportunity to register the GIC distribution along the 34° W longitude over several latitudes, i.e. 62°–69° N latitudes. The geographic coordinates of the substations are given in Table 3.1. The installed system has no analogy in Russia.

Each of the aforementioned nodes has two parallel power transformers (except the one on 330 kV Kondopoga substation) with the connection *wye*/Δ; therefore the

**Table 3.1**

**Geographical coordinates of the substations with installed GIC registration equipment**

| Name | Code | Latitude | Longitude |
|------|------|----------|-----------|
| Vikhodnoy | VCH | 68.83 | 33.08 |
| Revda | RVD | 68.63 | 33.25 |
| Lovozero | LOZ | 67.97 | 34.16 |
| Titan | TTN | 67.53 | 33.44 |
| Loychi | LCH | 66.08 | 33.12 |
| Kondopoga | KND | 62.22 | 34.36 |

measurement equipment is installed in one of the neutrals. The registration equipment is mounted in the grounded power transformers neutrals, which allows them to perform the GIC registration without causing any disturbance to the power transformer operation. The scheme of developed and installed measurement equipment is given in Fig. 3.3. The local system performs data collection, initial data processing, and the data transfer to the data collection center located at Polar Geophysical Institute, Apatity, on the hourly basis for further analysis. The data is proceeded in three stages: (1) data is imported in the database in its initial form; (2) data is selected from the database with its subsequent processing including filtration, decreasing signal frequency (from 10 Hz down to one-minute value), baseline correction, other; (3) data is converted into a format necessary for displaying the results on the internet [39].

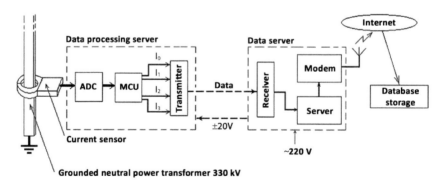

Figure 3.3: The GIC measurement equipment in the North-West Russia

Registered results have confirmed the preliminary theoretical assumption regarding high-level GIC distribution. The highest GICs are registered at the 330 kV substation Vikhodnoy, where 330 kV transmission transit suddenly changes its direction. The distribution of GIC amplitudes as a function of geomagnetic activity Kp at 330 kV substation Vikhodnoy is given in Fig. 3.4a, the frequency of maximum GIC appearance in Fig. 3.4b, and maximum GICs amplitudes per hour in Fig. 3.4c.

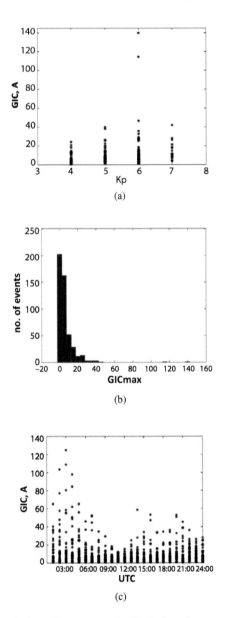

Figure 3.4: Characteristics of geomagnetically induced currents: (a) Distribution of geomagnetically induced currents amplitudes as a function of geomagnetic activity Kp; (b) Frequency of the maximum geomagnetically induced currents appearance; (c) Maximum geomagnetically induced currents amplitudes per hour [40]

Along with the more predictable GIC distribution, some extremes were registered which can pose threat to power grid operation. The list of them is given in Table 3.2 as a function of ionospheric source. The high GIC amplitude registered during GMD on June 29, 2013 is because of the fact that one of the parallel transformers was switched off. The largest so far registered GIC was on September 7, 2017 (Fig. 3.5).

**Table 3.2**

**Maximum geomagnetically induced currents values registered at 330 kV Vikhodnoy substation (adopted from [41])**

| Date | UTC | $I_{GIC_{max}}$, A | $dB/dt$, nT/min | Kp | Dst, nT | Source |
|------|-----|------|------|-----|---------|--------|
| 2017.09.07 | 23.30 | 85 | 1,000 | 8 | −142 | CME |
| 2016.09.02 | 01.08 | 80 | 495 | 6 | −60 | unknown |
| 2013.03.17 | 23.15 | 70 | 250 | 7 | −131 | CME |
| 2013.03.27 | 17.45 | 73 | 450 | 8 | −120 | CIR |
| 2013.06.29 | 02.10 | 140 | 810 | 6 | −100 | CME |
| 2015.03.17 | 23.15 | 60 | 220 | 7 | −220 | CME |
| 2015.09.08 | 01.35 | 48 | 625 | 6 | −60 | CIR |
| 2013.06.01 | 03.50 | 40 | 475 | 7 | −110 | CME |

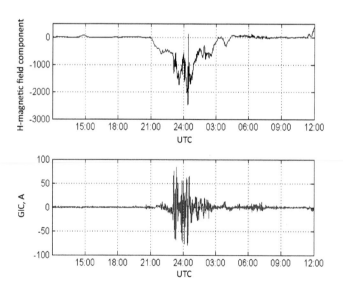

Figure 3.5: Upper panel: H-component of magnetic field at Lovozero; Bottom panel: geomagnetically induced currents at 330 kV Vikhodnoy substation during geomagnetic disturbance on September 7, 2017 [41]

The measurement results were used for evaluating which ionospheric current manifestation poses higher risks to the region. The importance of taking into account small-scale current structure field for GICs calculation, and extended ionospheric electrojet model is not sufficient [42], [43]. Activities towards the improvement of the GIC calculation model are also ongoing. The yearly-average correlation between GIC and variability of geomagnetic field components $|dX/dt|$ and $|dY/dt|$ was found to be rather high, $R \sim 0.7$, higher than that between GIC and magnetic perturbations $|\Delta X|$, $|\Delta Y|$, $0.5 < R < 0.7$ [44].

## 3.4 CONCLUSION

The final link in the space weather chain is the electric field excitation on the Earth's surface. In case any grounded conductive technological system is located in the affected region, GICs will be induced. The GIC amplitude can reach severe values leading to technological system operation disruption. The disruption degree can vary from minor to widespread long-lasting blackout. Detailed information on catastrophe models is given in Chapter 6.

There is no general rule determining how high should be the GIC amplitude in order to be considered as a destructive one. Experience shows that each technological system is susceptible to a unique set of GIC amplitudes. The relative comparison of three neighboring power grids located in Scandinavia showed that the highest GIC appearance does not certainly mean the highest danger to the power grid. Better understanding of critical values is the crucial task of the modern engineering.

Regular practice of GIC measurement is traditional for high-latitude countries, which experience GMD effects in the course of their evolution. The evident economic loss associated with these events boosted various branches of the GMD impact mitigation including the establishment of continuous GIC measurement sites. The growing awareness of negative GMD impacts results in an enlarged list of countries that observe the GIC. The GICs are measured in the several middle- and low-latitude countries including Brazil and South Africa.

Another question under focus is "which ionospheric conditions are more hazardous to terrestrial technological systems?". An improved understanding of the detailed connection between the ionospheric-magnetospheric conditions and their geomagnetic consequences is essential. This provides a solid basis for developing reliable warning and forecasting systems together. The accurate extreme value estimation is a silver bullet for optimal mitigation procedures development.

## REFERENCES

1. Narendra, K. (2016). Wide Area Real Time GIC Monitoring using TESLA Phasor Measurement Unit (PMU). *whitepaper, ERLPhase.* doi: 10.13140/RG.2.1.4024.2168.
2. Muchinapaya, A., Gaunt, C. T., Oyedokun, D. T. O. (2018). Design of a Low-Cost System to Monitor Geomagnetically Induced Currents in Transformer Neutrals. *2018 IEEE PES/IAS PowerAfrica*, 875–879. IEEE.

3. Campbell, W. H. (1980). Observation of electric currents in the Alaska oil pipeline resulting from auroral electrojet current sources. *Geophysical Journal International,* 61(2), 437–449. doi: 10.1111/j.1365?246X.1980.tb04325.x.

4. Viljanen, A., Koistinen, A., Pajunpää, K., Pirjola, R., Posio, P., Pulkkinen, A. (2010). Recordings of geomagnetically induced currents in the finnish natural gas pipeline–summary of an 11-year period. *Geophysica,* 46(1-2), 59–67.

5. Matandirotya, E., Cilliers, P. J., Van Zyl, R. R., Oyedokun, D. T., de Villiers, J. (2016). Differential magnetometer method applied to measurement of geomagnetically induced currents in Southern African power networks. *Space Weather,* 14(3), 221–232. doi: 10.1002/2015SW001289.

6. Matandirotya, E., Cilliers, P.J., van Zyl, R.R. (2015). Differential magnetometer method (DMM) for the measurement of geomagnetically induced currents (GIC) in a power line: technical aspects. *Journal of Electrical Engineering-Elektrotechnicky Casopis,* 66(7), 50–53.

7. Caraballo, R., Sánchez Bettucci, L., Tancredi, G. (2013). Geomagnetically induced currents in the Uruguayan high-voltage power grid. Geophysical *Journal International,* 195(2), 844–853. doi: 10.1093/gji/ggt293.

8. Marshall, R. A., Smith, E. A., Francis, M. J., Waters, C. L., Sciffer, M. D. (2011). A preliminary risk assessment of the Australian region power network to space weather. *Space Weather, 9(10).* doi: 10.1029/2011SW000685.

9. Vodyannikov, V. V., Gordienko, G. I., Nechaev, S. A., Sokolova, O. I., Khomutov, S. Y., Yakovets, A. F. (2006). Geomagnetically induced currents in power lines according to data on geomagnetic variations. *Geomagnetism and Aeronomy, 46(6),* 809–813. doi: 10.1134/S0016793206060168.

10. Torta, J. M., Marsal, S., Quintana, M. (2014). Assessing the hazard from geomagnetically induced currents to the entire high-voltage power network in Spain. *Earth, Planets and Space, 66(1),* 87. doi: 10.1186/1880-5981-66-87

11. Kalafatoğlu, E. C., Kaymaz, Z., Moral, A. C., Çağlar, R. (2015). Geomagnetically induced current (GIC) observations of geomagnetic storms in Turkey: Preliminary results. In *2015 7th International Conference on Recent Advances in Space Technologies (RAST),* 501–503. IEEE. doi: 10.1109/RAST.2015.7208396.

12. Ngwira, C. M., McKinnell, L. A., Cilliers, P. J., Viljanen, A., Pirjola, R. (2009). Limitations of the modeling of geomagnetically induced currents in the South African power network. *Space Weather, 7(10).* doi: 10.1029/2009SW000478.

13. Pirjola, R. (2005). Effects of space weather on high-latitude ground systems. *Advances in Space Research, 36(12),* 2231–2240. doi: 10.1016/j.asr.2003.04.074.

14. Viljanen, A., Pirjola, R. (1994). Geomagnetically induced currents in the Finnish high-voltage power system. *Surveys in Geophysics,* 15:383–408.

15. Pirjola, R. J., Viljanen, A. T., Pulkkineni, A. A. (2007). Research of geomagnetically induced currents (GIC) in Finland. In *2007 7th International Symposium on Electromagnetic Compatibility and Electromagnetic Ecology,* 269–272. IEEE.

16. Pirjola, R. J., Boteler, D. H. (2006). Geomagnetically induced currents in European high-voltage power systems. In *2006 Canadian Conference on Electrical and Computer Engineering,* 1263–1266. IEEE. doi: 10.1109/CCECE.2006.277540

17. Kappenman, J.G. (2005). An overview of the impulsive geomagnetic field disturbances and power grid impacts associated with the violent Sun-Earth connection events of 29–31 October 2003 and a comparative evaluation with other contemporary storms. *Space Weather, 3(8).* doi: 10.1029/2004SW000128.

18. Viljanen, A., Pulkkinen, A., Pirjola, R., Pajunpää, K., Posio, P., Koistinen, A. (2006). Recordings of geomagnetically induced currents and a nowcasting service of the Finnish natural gas pipeline system. *Space Weather, 4(10)*. doi: 10.1029/2006SW000234.

19. Clarke, E., Viljanen, A., Wintoft, P., et al. (2011). The EURISGIC database : a tool for GIC research. [Poster]. In *European Space Weather Week 8, Namur, Belguin, 28 Nov - 2 Dec.*, British Geological Survey.

20. Wik, M., Viljanen, A., Pirjola, R., et al. (2008). Calculation of geomagnetically induced currents in the 400 kV power grid in southern Sweden, *Space Weather, 6(7)*. doi: 10.1029/2007SW000343.

21. Torta, J. M., Serrano, L., Regué, J. R., Sanchez, A. M., Roldán, E. (2012). Geomagnetically induced currents in a power grid of northeastern Spain. *Space Weather, 10(6)*. doi: 10.1029/2012SW000793.

22. Lian-guang, L., Hao, Z., Chun-ming, L., Jian-hui, G., Qing-xiong, G. (2005, August). Technology of detecting GIC in power grids & its monitoring device. In *2005 IEEE/PES Transmission & Distribution Conference & Exposition: Asia and Pacific*, 1–5. IEEE. doi: 10.1109/TDC.2005.1546843.

23. Liu, C., Li, Y., Pirjola, R. (2014). Observations and modeling of GIC in the Chinese large-scale high-voltage power networks. *Journal of Space Weather and Space Climate, 4*, A03. doi: 10.1051/swsc/2013057.

24. Choi, K. C., Park, M. Y., Ryu, Y., Hong, Y., Yi, J. H., Park, S. W., Kim, J. H. (2015). Installation of induced current measurement systems in substations and analysis of GIC data during geomagnetic storms. *Journal of Astronomy and Space Sciences, 32(4)*, 427–434. doi: 10.5140/JASS.2015.32.4.427

25. Lesher, R. L., Porter, J. W., Byerly, R. T. (1994). SUNBURST/spl minus/a network of GIC monitoring systems. *IEEE Transactions on Power Delivery, 9(1)*, 128–137.

26. North American Electric Reliability Corporation (2011). Geo-magnetic Disturbances (GMD): Monitoring, Mitigation, and Next Steps. *A Literature Review and Summary of the April 2011 NERC GMD Workshop*. NERC, Atlanta.

27. Ngwira, C. M., Pulkkinen, A., McKinnell, L. A., Cilliers, P. J. (2008). Improved modeling of geomagnetically induced currents in the South African power network. *Space Weather, 6(11)*. doi: 10.1029/2008SW000408.

28. Koen, J., Gaunt, T. (2003). Geomagnetically induced currents in the Southern African electricity transmission network. In *2003 IEEE Bologna Power Tech Conference Proceedings, 1*, 7. IEEE. doi: 10.1109/PTC.2003.1304165.

29. Koen, J., Gaunt, C.T. (1999). Preliminary Investigation of GICs in the ESKOM Network. *Report to Eskom, University of Cape Town, December 1999*.

30. Trivedi, N. B., Vitorello, Í., Kabata, W., Dutra, S. L., Padilha, A. L., Bologna, M. S., Pirjola, R. (2007). Geomagnetically induced currents in an electric power transmission system at low latitudes in Brazil: A case study. *Space Weather, 5(4)*, 1–10. doi: 10.1029/2006SW000282.

31. Albert, D., Halbedl, T., Renner, H., Bailey, R. L., Achleitner, G. (2019). Geomagnetically induced currents and space weather - A review of current and future research in Austria. In *2019 54th International Universities Power Engineering Conference (UPEC)*, 1–6. IEEE. doi: 10.1109/UPEC.2019.8893515.

32. Marsal, S., Torta, J. M. (2019). Quantifying the performance of geomagnetically induced current models. *Space Weather, 17(7)*, 941–949. doi: 10.1029/2019SW002208.

33. Watari, S., Kunitake, M., Kitamura, K., Hori, T., Kikuchi, T., Shiokawa, K., Watanabe, Y. (2009). Measurements of geomagnetically induced current in a power grid in Hokkaido, Japan. *Space Weather, 7(3)*. doi: 10.1029/2008SW000417

34. Watari, S. (2015). Estimation of geomagnetically induced currents based on the measurement data of a transformer in a Japanese power network and geoelectric field observations. *Earth, Planets and Space, 67(1)*, 77. doi: 10.1186/s40623-015-0253-8

35. Pulkkinen, A., Bernabeu, E., Thomson, A., et al. (2017). Geomagnetically induced currents: Science, engineering, and applications readiness. *Space Weather*, 15:828–856.

36. Fisher, E., Eto, J. H., LaCommare, K. H. (2012). Understanding bulk power reliability: the importance of good data and a critical review of existing sources. In *2012 45th Hawaii International Conference on System Sciences*, 2159–2168. IEEE.

37. Sakharov, Y. A., Danilin, A. N., Ostafiychuk, R. M. (2007, June). Registration of GIC in power systems of the Kola Peninsula. In *2007 7th international symposium on electromagnetic compatibility and electromagnetic ecology*, 291–292. IEEE.

38. Viljanen, A. (2011). European project to improve models of geomagnetically induced currents. *Space Weather, 9(7)*.

39. Efimov, B., Selivanov, V., Sakharov, Y. (2019). Impact of Geomagnetically Induced Currents on Transformers in the Kola Power Grid. In *2019 International Multi-Conference on Industrial Engineering and Modern Technologies (FarEastCon)*, 1–4. IEEE.

40. Sakharov, Ya.A., Katkalov, J., Selivanov, V., et al. (2017). GIC registration in the regional power grid. In *XI Solar System Plasma Conference, Moscow*, 134–145 [in Russian].

41. Sakharov, Ya.A., Selivanov, V., Bilin, V.A., et al.., et al. (2019). Extreme geomagnetically induced currents values on the regional power grid. In *XLII Annual Seminar Physics of Auroral Phenomena, Apatity*. doi: 10.25702/KSC.2588-0039.2019.42.53-56 [in Russian].

42. Belakhovsky, V. B., Pilipenko, V. A., Sakharov, Y. A., Selivanov, V. N. (2018). Characteristics of the variability of a geomagnetic field for studying the impact of the magnetic storms and substorms on electrical energy systems. *Izvestiya, Physics of the Solid Earth, 54(1)*, 52–65.

43. Kozyreva, O., Pilipenko, V., Sokolova, E., Sakharov, Y., Epishkin, D. (2020). Geomagnetic and Telluric Field Variability as a Driver of Geomagnetically Induced Currents. In *Problems of Geocosmos–2018*, 297–307. Springer, Cham.

44. Vorobev, A. V., Pilipenko, V. A., Sakharov, Y. A., Selivanov, V. N. (2020). Statistical Properties of the Geomagnetic Field Variations and Geomagnetically Induced Currents. In *Problems of Geocosmos–2018*, 39–50. Springer, Cham.

# Part II

Inside the Power System

# 4 Reaction of Power Systems to Geomagnetic Disturbances

## CHAPTER CONTENTS

4.1 Geomagnetically induced current impact on power system equipment .......... 82
    4.1.1 Power transformers and autotransformers ........................................... 85
    4.1.2 Synchronous machines ........................................................................ 97
    4.1.3 Measurement transformers – Relay protection system .................... 102
    4.1.4 Geomagnetically induced current impact on other power system
        equipment ......................................................................................... 106
4.2 Modern approaches of geomagnetically induced currents integration into
    the power grid state calculation ................................................................. 109
4.3 Conclusion ................................................................................................... 115

POWER system equipment is designed to withstand normative disturbances corresponding to reasonable risk level. Risk level is chosen in accordance to production maintenance and replacement in case of unexpected loss. Majority of power system equipment is designed for a reliable operation within the 25 year interval. It is recommended to take into account the contingencies with the return period of 10 years. Modern industry does not consider negative GIC impact on power grid. GIC flows over power grid elements may violate power system state. GIC does not endanger power system operation conditions by themselves. The main challenge posed by GIC to power system state is the power system equipment operation conditions change provoked by GIC flow. Primarily, GIC affects grounded wye-connected power transformers and autotransformers by their core saturation. In turn, this brings indirect effects to the power system state. Contrary to more studied natural hazards such as earthquakes, windstorms, and floods, power system planners and operators have little experience of power grid operation whilst the GMD. The apparent correlation of power grid disturbance warrants the analysis and mitigation strategies implementation.

## 4.1 GEOMAGNETICALLY INDUCED CURRENT IMPACT ON POWER SYSTEM EQUIPMENT

The process of threat assessment is multi-phase. First, the threat should be character-ized. Then, the vulnerability assessment of the system's components is performed. Afterwards, the reaction on the system level is studied. Based on the achieved results, the system's restoration plan supported by relevant mitigation actions is developed.

Primarily, the scenario of GMD impact on power grid operation is determined by GMD size and energy hold. The GMD can be severe according to one parameter but moderate to others. The overview of GMD effects on power grid operation as a function of Kp and Dst indices is given in Table 4.1. Experience shows that the definition of extreme events is too broad. Extreme events can be subdivided into four categories based on the level of damage to power system equipment. **Slight** – dam-age stays for equipment tripping; **Moderate** – equipment malfunction or overheat-ing without equipment damage; **Strong** – overheating including equipment damage; **Catastrophic** – complete equipment destruction.

**Table 4.1**

**Geomagnetic storm severity according to the Kp and Dst indices**

| Scale | Kp | Dst | Effect |
|-------|----|-----|--------|
| Minor | 5 | $<-50$ nT | Weak power grid fluctuations may occur |
| Moderate | 6 | $-50-(-)100$ nT | High-latitude power systems may experi-ence voltage alarms, long-duration storms may cause transformer damage |
| Strong | 7 | $-100-(-)250$ nT | Voltage corrections may be required, false alarms triggered on some protection de-vices |
| Severe | 8 | $-250-(-)600$ nT | Possible widespread voltage control prob-lems and some protective systems will mistakenly trip out key assets from the grid |
| Extreme | 9 | $>-600$ nT | Widespread voltage control problems and protective system problems can occur, some grid systems may experience com-plete collapse or blackouts. Transformers may experience damage |

The GMDs impact the power grid operation by inducing low-frequency conduc-tive noise, so-called GICs. These currents are distributed over any grounded system of conductors (see Figs. 4.1 and 4.2).

GICs are considered to be a quasi direct currents (DC) since their frequency (around 1 mHz) is relatively low in comparison with nominal power system frequency (50/60 Hz). Moderate GICs can last from several minutes to several hours.

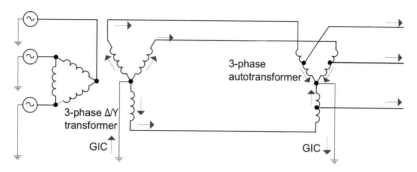

Figure 4.1: Principles of GIC distribution in power grid

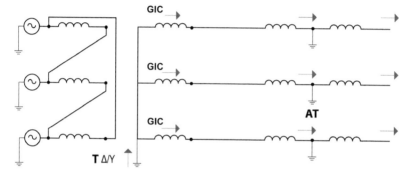

Figure 4.2: Principles of GIC distribution in power grid in case of the single-phase transformer group installation

High level GICs with a duration of a few minutes can happen among moderate GIC. Power system equipment is not designed to withstand long-lasting DC current flow. The avenue of primary GIC impact on power grid is through power transformers which are brought to half-cycle saturation. In case secondary electromagnetic effects are intensive, non-stationary electromechanical processes may occur, which result in grid's frequency drop/decrease, loss of dynamic and small-signal stability, widespread blackouts. Depending on the power transformer type installed in the system and the power grid parameters, the three key pathways to blackout are depicted in Fig. 4.3.

System level failures may range from no system impact at all to widespread outage with significant indirect losses. Halloween blackout in 2003 is characterized as a local power outage. Local power outage could also have damaged transformers as root cause. In that case, it may take a considerable time to return the system to normal operation and result in additional repair cost. Thus, we have the temporal load decrease or "unit at danger" disconnection in the moment of GMD appearance.

Power grid operators need to ensure satisfactory electrical energy quality and reliable power grid operation in the admissible continuous power system states. The

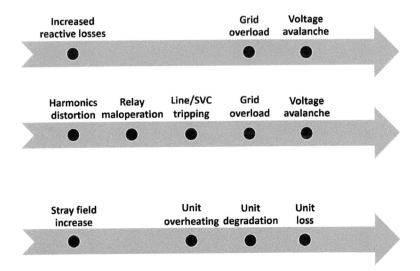

Figure 4.3: Pathways to blackout

power system state is characterized by a set of parameters: real ($P_g$) and reactive ($Q_g$) generation power, and loads ($P_l$, $Q_l$), power flows ($P_{ij}$, $Q_{ij}$), currents over network elements ($I_i$), voltage levels in the network nodes ($U_i$), frequency ($f_{grid}$). Operational limit constraints are given in inequalities (4.1–4.5)

$$P_{ij}^2 + Q_{ij}^2 \leq S_{ij_{max}}^2 \tag{4.1}$$

$$Ui_{min} \leq U_i \leq Ui_{max} \tag{4.2}$$

$$0 \leq I_i \leq Ii_{max} \tag{4.3}$$

$$Pg_{i_{min}} \leq f_{grid} \leq Pg_{i_{max}} \tag{4.4}$$

$$Qg_{i_{min}} \leq f_{grid} \leq Qg_{i_{max}} \tag{4.5}$$

To ensure the reasonable level of power supply reliability, the set of values **M** ($U$, $I$, $P$, etc.) is regulated for the most typical power system states. The power system state is characterized by the new set of values **M'**, part of which can exceed the operational limits. Along with the power system equipment technical characteristics, the power grid architecture also determines the maximum admissible limits of power system parameters fluctuation.

The analysis of secondary electromagnetic and thermal effects from GIC is based on the comparison of two power system states, i.e. before and after the contingency.

The admissible values of the power system parameters are determined in accordance with the current legal norms. Further, the detailed information about the GIC's impact on power system equipment is given.

### 4.1.1  POWER TRANSFORMERS AND AUTOTRANSFORMERS

Power transformers are electrical devices installed on the power plants or substations and are employed in the power network to increase or decrease the voltage level with the same frequency. Normally, the total installed power transformer capacity in the grid is four-five times higher than the generation capacity. Different power transformer's configurations exist, though the most popular one is the three-phase power transformer. Compared to a single-phase power transformer, the power losses are 12–15% lower in three-phase power transformers and their production cost is reduced by 20–25%. The choice of power transformer type is a multifactorial process which is defined by the power transformer weight, size and transportation constraints apart from electromagnetic characteristics. Therefore, single-phase transformers are mainly used as ultra-high voltage grid transformer or the step-up transformer on the high-capacity power plant.

The reliability requirements to a power transformer operation are determined by the degree of its influence on the whole power grid security. Particularly, high requirements are associated with ultra-high voltage grid transformer or the step-up transformer on the high-capacity power plant. The power transformer design should consider the low-probability contingencies (once in ten years). GMDs accounting is not a regular practice.

Transformer replacement is costly and logistically challenging. Table 4.2 gives estimates of the recovery time for power transformer failures conjectured from the information about the response to natural hazards and supplemented with information from the technical documentation [1]. US Department of Energy estimates the average lead time of a domestically manufactured transformer between 5–12 months and internationally manufactured 6–16 months and can be up to 28–24 months in high demand periods [2]. The Royal Academy of Engineering, UK, says that it will take at least eight weeks to transport, install and commission a spare transformer unit [3]. Supply is also hampered by a surge in demand from India, China, Latin America and Middle East, where vast new grids are being constructed to cope with the increased demand for electricity power [4].

Real case of power transformer failures and corresponding repair time include following examples [5]. Transformer fire at Krümmel Power Station, Germany, in June 2007 did not result in the unit loss, and the transformer was brought back to operation by circuit switching in 30 minutes after the event. Time for onsite repair varies from 48 hours (case of power transformer damage at Rostov Nuclear Power Plant, Russia, in October 2010) to 960 hours (case of 1,200 MVA step-up transformer failure at Salem Nuclear Power Plant, US, March 1989). In case of no replacement is available on site, the repair time rose up to 6,480 hours for the case of generator transformer failure at Longannet Power Plant, Scotland, in September 2009. The chronic of power transformer repair after Hydro-Québec event in 1989 was the following.

**Table 4.2**

**Power transformer repair time as a function of its type and the damage severity**

| Repair strategy | Repair time |
| --- | --- |
| Inspect, reset and re-energize | 14–20 hours |
| Refill oil, onsite | 2 days |
| Minor repair onsite | 1–2 weeks |
| Change windings, onsite | 3 months |
| Replace (with spare) | 5 days |
| Replace (no existing spare) | 1 year |

Transformer T1 (phase C) at LG-4 was the first one to put back into service on April 26, 1989. Power transformer T3 at LG-4 went back to operation on December 15, 1989. Static VAR compensator CLC 12 at Chibougamau was replaced by a spare unit on June 1, 1989 and went back into service in July 1990. The sequence of events that lead to a blackout is given in Table 2.2.

Modern high-voltage power transformers have a complex structure. The main components include: tank, core, windings, isolation, tap changer, cooling system, leads and terminal arrangements. Core is the "heart" of the power transformer, its constructive basis. It creates a closed magnetic circuit with low reluctance for carrying the linkage flux through the windings. The core is stacked by a lamination of thin (ca. $\leq 0.3$ mm) electrical steel sheets, which are coated with a layer of insulation material (thickness ca. 10–20 $\mu$m). Core materials are constantly improved by introducing new materials such as oriented, hot-rolled grain-oriented (HRGO), cold-rolled grain-oriented (CRGO), high permeability cold-rolled grain-oriented (Hi-B), laser scribed steels [6]. The core consists of limbs (**L**) surrounded by windings and connecting yokes (**Y**). The core joint is the place where limbs and yokes are met to each other. The most common core joint is a mitered joint with a 45° cutting angle. Based on the number of steps, two types of core joints are considered: overlap and step-lap (more than one step).

High-voltage power transformer windings can be layer or disk. In the first case, both high-voltage and low-voltage transformer windings are made in cylindrical shape and placed concentric relative to each other. This method gives the best utilization of space and is widely used in power transformer design, however, it demands a lot of time and labour, and involves high cost and production time. In the second case, the windings are made as cylinders with the same diameters and placed above each other on the same limb. Windings are made from Cu transposed wires.

From the construction type of the core and windings, the power transformer can be classified into core-form and shell-form [7]. The graphical representation of them is given in Fig. 4.4. The windings are wrapped around the core with a cylindrical shape in core-form transformers, however, the core is stacked around the windings in shell-form transformers. Each winding consists of two parts (1 and 2). Windings have a flat or oval shape in this case and are often called pancake windings. The total cost of active materials is 20–30% less for core-type transformer, which makes this type more preferable for medium- and high-voltage levels.

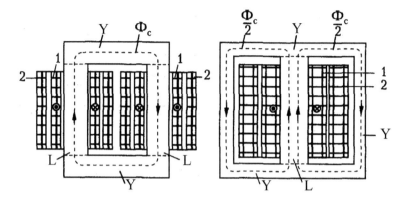

Figure 4.4: Power transformer construction schemes: core-form transformer on the left and shell-form transformer on the right (Courtesy of Prof. Popov, V.)

Single-phase power transformers can be made with two, three or four limbs. In case of two limbs, there is not return limb and the cross-section of the yoke is the same as for the limb. In case of three or four limbs, there are two return limbs. The return limbs are not surrounded by any windings and constructed in order to provide a closed path for the flux. The size of three- and four-limb transformers is decreased compared to two-limb one, since the cross-section of the yokes and return limbs is ca. 50% of the main limb. Three-phase transformers have three or five limbs. In the first case, all three limbs are surrounded by windings. The five-limb design is used for large power transformers to decrease their height, stray fluxes and eddy current losses.

The power transformer's active part is located in the tank filled with the insulation oil. It is made in a rectangular cubic shape from soft magnetic steel with a thickness about few centimeters. The tank should be robust to mechanic, acoustic thermal, electric and electromechanical contingencies and sustain transportation. The tank should minimize the stray fields outside the transformer and eddy currents inside the tank. The shielding is used in large power transformers. The shielding is done by mounting the laminations of copper or high permeable material on the inner side of the tank wall that are so-called protective shunt [6].

Autotransformers are types of transformers in which the primary and secondary windings have a common part and they are electrically connected at that common point [8]. In other words, power is transmitted using both galvanic and inductive methods in autotransformers. They are characterized by a reduced size and an optimum use of construction materials together with low short-circuit impedance that prevents their widespread installation in the grid. Generally, they are used when the voltage ratio is not so big.

The power transformer passport data includes the following parameters:

1. Rated power is the value of the total power indicated in the passport, at which the transformer can be continuously loaded under the rated conditions of the installation site and the cooling conditions at the rated frequency and voltage. The rated power of a two-winding power transformer is the power of each of the windings. The windings of the three-winding transformer can be made for the same power or for the different. In the last case, the rated power is the power of one of the windings.
2. Rated voltage is the voltage of the first and secondary transformer windings in the no-load condition. This rated voltage of the three-phase transformer is equal to its line-voltage, or is equal to phase voltage ($U_{line}/\sqrt{3}$) in case of a single-phase transformer. The transformation coefficient is specified for each pare of windings in three-winding transformer.
3. Short-circuit voltage $u_k$ is the voltage of one of the windings when the current flowing over the winding is equal to the rated one whilst the other is short-circuited. Short-circuit voltage defines the windings impedance. In case of three-winding power transformer, the test is done for each pair of the windings. Short-circuit voltage is defined in the percentages of the rated voltage, and is in the range of 2–10%.
4. No-load current defines active and reactive losses in the transformer's steel and depends on the steel's magnetic properties, power transformer design and production quality. No-load current is defined in the percentage of the rated current, and normally does not exceed the value of 2%.
5. No-load losses consists of the hysteresis losses (ca. 50–80% of no-load losses) and eddy current losses (ca. 20–50% of no-load losses). Hysteresis losses are caused by the frictional movement of magnetic domains in the core laminations being magnetized and demagnetized by alternation of the magnetic field. Hysteresis losses depend on the core steel.
6. Short-circuit losses consists of Ohmic heat losses (copper losses) and conductor eddy current losses. The Ohmic heat losses is determined by the power transformer total load and is caused by the resistance of the conductor. It can be reduced by increasing the cross-sectional area of conductor or by reducing the winding length.
7. Connection scheme. In total, three connection scheme exist: delta ($\Delta$), star (*wye*), grounded star (*wye*$_0$). The graphs are given in Section 2.4.3. $Y$ connection allows us to design the inner isolation by considering the phase electromotive force which is $\sqrt{3}$ times lower than the linear electromotive force. Star connection is normally used for high-voltage winding. Delta winding is generally used for low-voltage winding, since it allows us to reduce the conductors cross-sectional area by considering phase current ($\sqrt{3}$ times lower than the linear current). Moreover, delta connection prevents high harmonics (with the numbers $n = 3k$, $K = 1, 2, 3, ...$) penetration to the grid.

8. Transformer windings group specifies the angle between the electromotive forces of the first and the second windings. Twelve groups exist. In case both of the windings are wye-connected, even group can be achieved 2, 4, 6, 8, 10, 0. In case the combination of $\Delta/\Delta$ or $\Delta/wye$ windings are used, the odd group can be achieved 1, 3, 5, 7, 9, 11. The group and connection scheme are specified as following $wye/\Delta - 11$.

The whole set of power transformer parameters is rarely available. Parameters which differ from those specified above can be requested to the manufacturer. Some parameters can be obtained from terminal measurements. For instance, leakage inductance and winding resistances can be obtained when the high voltage winding is energized and low voltage winding is short circuited. Power transformers of types described above can be constructed for any rated power or voltage level, and from any magnetic material. Moreover, the construction type does not influence the functional principle, electromagnetic processes or the ability to ensure required outputs.

The GMD poses several ill effects to power transformer operation. In order to understand what happens at the moment of GIC appearance, a saturation simplified magnetization curve is given in Fig. 4.5. Transformers are designed to operate under the sinusoidal voltage and current. GIC as a quasi DC imposes a unidirectional flux in the transformer's core. In turn, the DC flux adds to the AC flux in one half-cycle and subtracts from the AC flux in the other half. In other words, magnetization current becomes very high in one half-cycle and decreases a little in the other half-cycle. If large enough, GIC can result in power transformer's half-cycle saturation. Since the high-voltage power transformers have a big number of windings turns, even small GIC may saturate the core. Whenever the iron core saturates, its relative permeability tends to decrease. In case of a deep saturation, permeability is equal to the one of the air $\mu_r = 1$. The improved saturation curve for single-phase transformer which enables a consistent depiction of real BH data is given in [9].

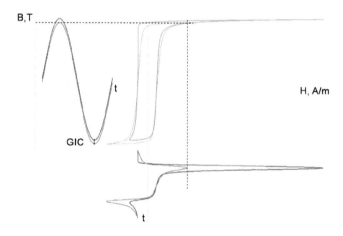

Figure 4.5: Magnetic flux curve shape

The magnetic properties of electrical steel are usually measured with a sinusoidal magnetic flux density at a fixed level of magnetic induction [10]. Hi-B electrical steel type materials with higher saturation flux densities, lower permanent flux values, and larger linear portion of magnetic curve compared to regular grain-oriented (RGO) materials are used for power transformer's core. The numerical representation of saturation core for a single-phase four-limb power transformer is shown in [11]. As the transformer core saturates, magnetic flux looks for new paths with relatively low reluctance, e.g. structural components of the transformer made of the ferromagnetic material. As these parts are not designed to minimize eddy currents, induction from the leakage flux heats up the elements. The response of power transformer to GIC as function of its core design was in the scope of various studies [12], [13], [14], [15], [16].

According to the principle of duality between electric and magnetic circuits, each flux path can be represented as an inductor [17]. Magnetic flux paths differ in different transformer designs. The generalized magnetic circuit is shown in Fig. 4.6. The magnetic circuit of a single-phase transformer is represented with two side branches. Three middle branches correspond to the three-phase three-limb transformer magnetic circuit. The full circuit in Fig. 4.6 constitutes to three-phase five-limb transformer magnetic circuit.

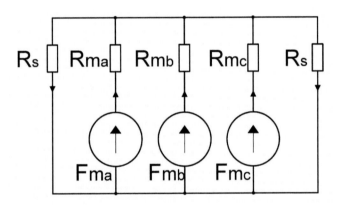

Figure 4.6: Generalized power transformer magnetic circuit

The DC flux path, connection scheme, and insulation characteristics determine power transformer's robustness to GMD effects. The analysis result shows that the single-phase transformers are the most vulnerable regardless to the number of limbs. In contrast, three-phase three-limb is the most robust and reacts with minor asymmetries, since DC fluxes compensate each other in the main limbs and connecting yokes. In other words, the paths of the positive and zero sequence fluxes are different, where positive and zero sequence fluxes are the representation of the physical fluxes in the transformer's core in the sequence domain 0-1-2, following the rule that any set of unbalanced three-phase quantities can be represented as the sum of three symmetrical set of balanced phasors. The positive sequence fluxes are the normal power system phase sequence quantities whereas zero sequence fluxes are a set of three phasors with similar phase information. The zero sequence path is closed through an air gap with high magnetic reluctance. The positive and zero sequence flux paths are both closed within the core in a single-phase transformer, which is the most vulnerable type. Three-phase five-limb power transformers fall between these two extremes. Although both direct and zero sequence fluxes are closed inside the core, they have different paths. The return branches are the paths for the zero sequence fluxes. In case the magnetic tank of a three-phase three-limb transformer is designed in a way that an air distance to a core is small, there is a chance that the tank acts as a path for zero-sequence fluxes. Hence, the limbs of three-limb core saturate, since the zero sequence reluctances are reduced. The relative power transformer susceptibility is given in [18]. The flux distribution in power transformers core due to GIC is presented in Fig. 4.7a–4.7c [19]. The correlation between the transformers magnetization current ($I_m$) and GIC ($I_{GIC}$) can be described as follows (Eq. 4.6):

$$I_m = \frac{\pi\,(1 - cos\alpha)}{sin\alpha - \alpha cos\alpha} I_{GIC} \tag{4.6}$$

where $\alpha$ is saturation angle. Saturation angle is the angle corresponding to the time duration when the core flux exceeds the knee point to reach its maximum in case one considers a voltage cycle as $360°$.

The sensitivity of different transformer types to the effects of GIC's is sometimes compared and ranked according to the iron area which is available to the DC-flux generated by the windings [20]. A value of 0 stays for the most robust and a value of 1 stays for the most sensitive construction. The results are presented in Table 4.3.

**Table 4.3**
**Power transformer's sensitivity to GIC as a function of construction scheme**

| Transformer type | Number of phases | Number of legs | GIC sensitivity, p.u. |
| --- | --- | --- | --- |
| Full-wound, core-form | 3 | 3 | 0 |
| Full-wound, core-form | 3 | 5 | 0.24–0.33 |
| Autotransformer, shell-form | 3 | 3 | 0.5–0.67 |
| Full-wound, shell-form | 1/3 | 2/7 | 1 |
| Autotransformer, shell-form | 1 | 2/3 | 1 |

The winding connection scheme influences the power transformer robustness to GICs. In case the GICs are applied to a $Wye/\Delta$ power transformer, the following current and magnetic fluxes distribution are observed. The harmonics with the numbers $n = 3k$, $K = 1,2,3,...$ do not exist. Therefore, magnetic flux has highly non-sinusoidal character and contains high harmonics together with the main one (Fig. 4.8). High harmonics of magnetic flux $\Phi_Y$ generate electromotive difference

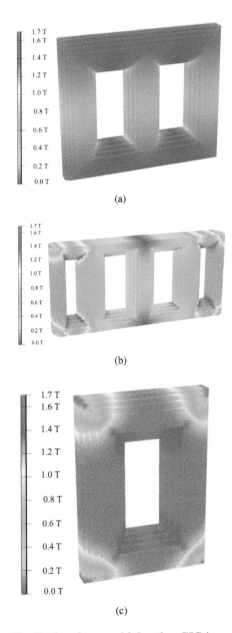

(a)

(b)

(c)

Figure 4.7: (a) Flux distribution due to a high value GIC in a core of a three-phase three-limb power transformer; (b) Flux distribution due to a low level GIC in a core of a three-phase five-limb power transformer ; (c) Flux distribution due to a very low level GIC in a core of a single-phase two-limb power transformer [19]

of equal amplitude and coincident in phase in the $\Delta$ connected winding. These electromotive difference induce inductive currents. Fluxes induced by these currents $\Phi_\Delta$ almost fully compensate the $\Phi_Y$ fluxes. Therefore, the resulting flux is almost sinusoidal.

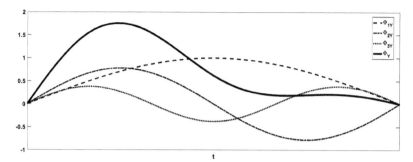

Figure 4.8: Magnetic flux curve shape

In case GICs are applied to $\Delta$ connected winding, the harmonics with the numbers $n = 3k$, $K = 1,2,3,...$ do not exist in linear currents. Currents of these harmonics circulate within the closed triangular. Magnetic fluxes of the harmonics with the numbers $n = 1k$, $n = 2k$ $(K = 1,2,3,...)$ of the primary are almost fully compensated by the fluxes from the secondary winding. Therefore, implementation of $\Delta$ winding is a practical GIC mitigation measure.

The power transformer's half-cycle asymmetrical saturation also leads to high harmonics generation. High harmonics are recognized as a drawback because they result in a sudden increase in the load and variation. Also they lead to the distortion of current and voltage waveforms that causes, as an example, thermal stress of equipment or unwanted tripping of protection circuit components. The specific feature of the generated spectrum is its richness in even and odd harmonics with low order harmonics reaching high levels [21]. In contrary, harmonics generated by FACTS, SSSCs, HVDC equipment and non-linear loads have the prevalence of odd harmonics. The value of total harmonic distortion (THD) is chosen in accordance with power system conditions (Eq. 4.7). The THD cannot exceed the following value THD$< 0.05$ [22]. THD is defined as in Eq. 4.7. The harmonic spectrum ranges normalized to the fundamental harmonic for three cases: normative distribution, single-phase power transformer saturation, three-phase five-limb power transformer saturation are given in Table 4.4. The values given in Table 4.4 are the average value for different GIC levels based on literature research. It is shown in [19] that damping of high harmonics at high GIC level occurs faster for all the transformer types.

$$THD = \frac{\sqrt{I_2^2 + ... + I_i^2 + ... + I_n^2}}{I_1}, \qquad (4.7)$$

where $I_i$ – is a relative amplitude of $i$ current harmonic and $I_1$ – is an amplitude of a main current harmonic.

**Table 4.4**
**High harmonic distribution in case of GIC appearance**

| Harmonic order | IEC 61000-4-7 | Single-phase | Three-phase five-limb |
|---|---|---|---|
| 1 | 1 | 1 | 1 |
| 2 | 0.005 | 0.8 | 0.35 |
| 3 | 0.015 | 0.32 | 0.15 |
| 4 | 0.003 | 0.12 | 0.07 |
| 5 | 0.015 | 0.05 | 0.05 |
| 6 | 0.002 | 0.09 | 0.03 |
| 7 | 0.01 | 0.05 | 0.01 |
| 8 | 0.002 | 0.01 | 0.01 |

GMDs may lead to an immediate power transformer outage or its postponed loss due to its accelerated insulation degradation. Following accidents are already registered. The GSU power transformer loss at PSE& G in NJ, US, during the March, 19, 1989 event is the example of an immediate transformer loss [23]. The shell-form transformer with an old winding lead design that made it susceptible to overheating caused by high circulating current was taken out of service a week later because of significant gassing [24]. A dissolved gas-to-oil detector showed an increase of 50 ppm. Visual inspection of the failed transformers revealed a severe damage to one of the two long series connections of the outer low-voltage winding paths. Phases **A** and **C** had a 20–25% conductor damage and phase **B** experienced insulation discoloration. In terms of power grid operation, this power transformer loss is separated from Hydro-Québec blackout. It also noted that within two years after 1989 event GSU transforms experienced failures at 11 nuclear power plants [23]. Photos of the damaged transformer are shown in many publications to illustrate consequences of GICs [25]. Nevertheless, there was no direct measurement of magnetic and thermal effects inside transformers during real GIC events so far.

The main danger posed by GICs to power transformers located in the "low-risk" regions is a number of partial discharge growth. Partial discharge stays for an electrical discharge or spark that bridges a small portion of an insulation between two conducting electrodes. It was reported that the series of geomagnetic events in 2003/4 caused significant winding overheating in a few large core-form power transformer in South Africa [26]. These incidents were found to coincide with winding overheating attributed to high sulphur content in some types of transformer's oil. Another example is the Transformer T4 at Halfway, New Zealand, failure one minute after the GIC occurrence. It does not seem possible for a transformer of this sort to fail this quickly due to saturation of the core creating hotspots. However, deterioration of transformers is cumulative and caused by events such as power system faults, electrical overloading GICs, etc. which over time degrades the transformer. It is possible that the transformer was already prone to failure and the GIC was the final contributor, hence failing very soon after the event [27].

The transformer health is directly related to long-, medium-, and short-term condition of the insulation [28]. Ambient temperature is an important factor in determining the load capability of a transformer [29]. The document also notes that the performance may be affected by the higher operation temperature. Many long-term aging processes originate in oil and insulating paper. Paper condition is affected by oxygen levels and high temperature, with degree of polymerization being an indicator of paper degradation [30]. It reduces the heat transfer and accelerates degradation. Common approach of measuring the top oil and winding temperatures can not access delayed GIC impact. Therefore, low energy degradation triangle method is proposed in [31]. The three components of the triangle are the dissolved gasses $H_2$, $CH_4$, $CO$, each measured in parts per million. The explanation of power transformer insulation degradation during 2003/4 events is given in Fig. 4.9.

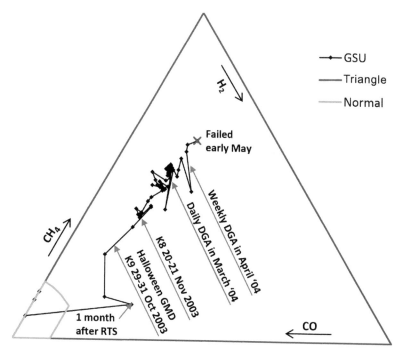

Figure 4.9: Low energy insulation degradation triangle for power transformer during 2003/4 geomagnetic events (Courtesy of Gaunt, C.T.)

The field tests of power transformer behavior under GIC were performed. In the first case, the 400 kV 400/400/125 MVA full-wound three-phase power transformer with a five-limb iron core was subjected to DC [33]. The stepwise neutral current increase from 50 A to 200 A was made. The highest temperature (ca. 130 °C) was recorded on the inside of a top yoke clamp, though the largest temperature rise was seen on the inside of a bottom yoke clamp. The highest temperature gradient, inside the bottom yoke clamp to surrounding bottom oil, was about 110 K. The time constant was 10 min. The test was done with the low ambient temperature $-2\,^\circ$C, hence, the temperature of 170–180 $^\circ$C could be achieved with the regular ambient temperature conditions. This can pose risk to old transformers and result in some dangerous gassing in an oil.

The "effective GIC" which considers different GIC values in the series and common windings of an autotransformer is under the scope while assessing GMD impact on power transformer operation. The term was introduced by [34]. The application of "effective GIC" for two-winding and three-winding power transformer modeling was adopted by [35] and [36] respectively. The "effective GIC" is the GIC that flows from the high voltage level (HVW) to ground and produces the same magnetic flux in the core as the combination of GIC in all the windings. In other words, effective GIC is the effective per phase current. Equations for equivalent flux and effective GIC for different power transformer types is given in Table 4.5. In general, the "effective GIC" equation has the common structure for which the winding current is proportional to the number of turns of the windings and inverse to the number of turns of the windings in which the effective GIC is considered to flow.

**Table 4.5**
**List of equations for effective GIC determination**

| Transformer type | Equivalent flux | Effective GIC |
|---|---|---|
| Two-winding transformer | $I_1 N_1 + I_2 N_2 =$ $I_{eff} N_1$ | $I_{eff} = I_1 + I_2 \frac{N_2}{N_1}$ |
| Three-winding transformer | | $I_{eff} = I_1 + I_2 \frac{N_2}{N_1} + I_3 \frac{N_3}{N_1}$ |
| Two-winding autotransformer | $I_s N_s + I_c N_c =$ $I_{eff}(N_s + N_c)$ | $I_{eff} = I_s \frac{N_s}{N_s+N_c} + I_c \frac{N_c}{N_s+N_c}$ |
| Three-winding autotransformer | | $I_{eff} =$ $I_s \frac{N_s}{N_s+N_c} + I_c \frac{N_c}{N_s+N_c} + I_3 \frac{N_3}{N_s+N_c}$ |

The proper information on power transformer's susceptibility to GMD helps to develop optimal mitigation strategies (see Chapter 5). Just the knowledge on GIC amplitude cannot justify "pass" or "suffer the damage". [37] proposes to divide power transformers in four categories as a function to their technical-geographical vulnerability to GMD such as *Category I* transformers not susceptible to GIC effects, *Category II* transformers least susceptible to core saturation, *Category III* transformers susceptible to core saturation but only some overheating of windings and structural parts registered, *Category IV* transformers susceptible to core saturation as well as possible damaging windings appearance or structural parts overheating registered.

These four categories define designed-based power transformers susceptibility to GIC which refers to total GIC susceptibility by considering actual GIC amplitude at a certain node.

## 4.1.2 SYNCHRONOUS MACHINES

Synchronous machines are another critical power system equipment type, which is characterized by a high cost and long production period. Synchronous machines are commonly used as generators especially for large power grids, such as turbine generators and hydroelectric generators in the grid power supply. Consequently, tight reliability requirements are applicable to its design and maintenance conditions. Their technical characteristics are determined by the power grid needs together with the industry patentability. The intense growth of the synchronous machine specific power happened in the 50s–80s of the twentieth century which was provoked by the development of the modern high-voltage power grids. It was caused by the following reasons:

1. Production of a generator with the higher specific power reduces the cost per MW (Table 4.6).
2. Production cost decrease.
3. Specific metal content of a generator decrease.
4. Total power plant operation cost decrease as well as reduced fuel consumption.
5. It helps to achieve higher power plant efficiency, since the total energy conversion efficiency of several generators $\eta_n$ is lower than the energy conversion efficiency of a one generator $\eta$ with the power equal to the sum of the smaller generators powers.
6. Change of generators working conditions and their efficiency decrease ($\eta > \eta_n = \eta_1 \cdot \eta_2 \cdot \eta_n$)

---

**Table 4.6**

**The comparison of turbogenerator characteristics as a percentage of the same characteristics for 60 MW model**

| Power, MW | Specific cost per MW, % | Electric energy production cost, % |
|---|---|---|
| 60 | 100.0 | 100.0 |
| 120 | 96.5 | 88.5 |
| 200 | 81.5 | 70.0 |
| 350 | 75.0 | 68.2 |
| 500 | 69.5 | 65.0 |

---

The power increase is limited by the ratio shown in Eq. 4.8. On the world level, the tendency to increase the specific synchronous machine power remained regardless the existing trend to construct decentralized power grids. The driving factors are the following: several countries remain interested in the construction of the power high-voltage power grids, the need to change the aged equipment even in the countries who are focused on the decentralized generation.

$$\frac{\text{production cost} + \text{operation cost}}{\text{replacement cost} + \text{electricity undersupply losses}} \qquad (4.8)$$

All synchronous machines can be divided into two big classes: the magnetic poles can be either salient or non-salient construction (hydro- and turbogenerators). Contrary to turbogenerators which are represented in the series of a mass production, hydrogenerators are designed as a unique piece for each hydro power plant. Therefore, their replacement in case of an emergency loss involves higher cost and reparation time. Within the classes synchronous machines differ in the specific power, construction and cooling system types. Cooling system types include air, water, hydrogen and water-hydrogen cooling systems. Transition to more efficient cooling system types allows to reduce the synchronous machine parameters such as dimensions and mass. For instance, the transition for the 100 MW turbogenerator from the indirect air cooling system of stator and rotor windings and direct air cooling system of magnetic core to indirect hydrogen cooling system of all active synchronous machine parts helped to decrease its total mass by 1.6 times and its active length by 2 times. Moreover, the cooling system improvement results in more rational usage of the active and construction materials.

Similarly to power transformers, synchronous machines are operated in parallel. Synchronous machines parameters are chosen in accordance to admissible thermal and electromagnetic forces. Nevertheless, the GIC impact on their operation is not considered currently. It is mentioned in [38], [39] that severe GMDs may result in overheat of rotor ending wings and give rise to mechanical vibrations. The real picture is more complicated.

The direct GIC impact on synchronous machines operation is associated with GIC flow over stator and rotor windings. It is assumed that induced GIC values are equal in all three phases, since synchronous machine size is infinitesimal small in comparison to a size of a geographical area affected by GMD. In this case, the field induced in stator windings does not induce any field in rotor windings. Thereby, synchronous machines experience only electromagnetic, thermal, and mechanical impact on stator windings.

On March 21, 1991 the GIC equal to 201 A (as a 1-minute mean value) was registered at 400 kV substation Rauma in the Finish power grid [40]. It is one of the highest ever registered GIC values which are published in the open source data. The field induced by this current is ca. 0.005 T, which is incomparably small to the nominal flux density of synchronous machine – 0.9 T. Thereby, the field induced by GIC is too small to cause any stator winding degradation.

It is stated that the admissible synchronous machine power $P_{adm}$ has to be reduced while operating with unacceptable prevalence of high harmonics. THD is equal to 0.92 for harmonic spectrum in case of single-phase transformer saturation and is equal to 0.39 in case of three-phase five-limb power transformer saturation.

High harmonics induce electromotive forces and currents in the stator windings, which cause additional losses in steel's active part, overheating, vibration and noise. These induced currents have the following frequency $f_{st} = f_{grid} \cdot N$, where $N$ – is the high harmonic number. Among them one distinguishes high harmonics of the range $N_1 = 1; 4; 7; ....; (3k-2)$, high harmonics of the range $N_2 = 2; 5; ....; (3k-1)$, and high harmonics of the range $N_3 = 3; 4; ....; (3k)$, where $k$ – is any positive natural number ($k = 1; 2; ....$). Harmonics of the frequency $N_1$ create rotating moment and multitude of harmonics $N_2$ create breaking moment. The set of high harmonics $N_3 = 3k$ create vibration moment.

The admissible power $P_{adm}$ is formed by the admissible power of stator windings $P_{adm,st}$ (losses in the Cu) and by admissible power due to high harmonics in the active steel of the stator and rotor windings $P_{adm,steel}$. In details, the methodology of $P_{adm}$ calculation is described in [41], [42].

Losses in stator windings caused by high harmonics can be found as follows. The Ohmic loss $Q$ due to high harmonics are found as a sum of Ohmic losses from each harmonic (Eq. 4.9):

$$\sum Q_{ohm} = Q_{ohm,1} + ... + Q_{ohm,i} + ... + Q_{ohm,n} \qquad (4.9)$$

The mathematical statement for Ohmic losses of $i$-harmonics is as follows (Eq. 4.10):

$$Q_{ohm,i} = \frac{1}{2} m_f (K_i I)^2 R_f, \qquad (4.10)$$

where $m_f$ – is number of phases; $K_i$ – is harmonic coefficient; $I$ – is current amplitude; $R_f$ – stator winding resistance at the temperature 75 °C.

The additional losses $Q_{add}$ are described as follows (Eq. 4.11):

$$\sum \Delta Q_{add} = \Delta Q_{add,1} + ... + \Delta Q_{add,i} + ... + \Delta Q_{add,n} \qquad (4.11)$$

The mathematical statement for additional losses of $i$-harmonic is (Eq. 4.12):

$$\Delta Q_{add,i} = \frac{1}{2} m_f (K_i I)^2 R_f \Delta K_{F,i}, \qquad (4.12)$$

where $\Delta K_F$ is a coefficient which states for additional losses caused by skin effect in stator windings and depends on the field frequency $\Delta K_{F,i} = \Delta K_{F,1} N^2$.

By taking into account Eqs. 4.9 and 4.11, the total losses in stator windings caused by high harmonics $Q^*$ can be described as in Eq. 4.13:

$$Q^* = Q_{ohm} \cdot \Delta Q_{add} = \frac{1}{2} m_f I^2 R_f (1 + \Delta K_{F,1}) \qquad (4.13)$$

Equation (4.14) determines the admissible power of stator windings $P_{adm,st}$ limited by its overheating.

$$P_{adm,st} = \sqrt{\frac{1 + \Delta K_{F,1}}{\sum Q_{ohm} + \sum \Delta Q_{add}}} \qquad (4.14)$$

The losses caused by high harmonics in the active steel of stator and rotor $P_{adm,steel}$ also limit admissible power $P_{adm}$ of synchronous machine. They can be calculated as in Eq. 4.15:

$$P_{adm,steel} = \frac{1}{\sqrt{\sum W_{N_{steel}}}}, \qquad (4.15)$$

where $\sum W_{N_{steel}} = 1 + \ldots + (\frac{s_i K_i}{s_1 K_1})^2 (i)^{1.3} + \ldots + (\frac{s_n K_n}{s_1 K_1})^2 (n)^{1.3}$; $s_i$ – screening coefficient, which is defined as $s_i = \frac{F_i^{res}}{k_{sat,i}F_i^{stator}}$, where $F_i^{res}$ – magneto motive force in air gap at the frequency $i$, $F_i^{stator}$ – stator magneto motive force at the frequency $i$, $k_{sat,i}$ – saturation coefficient at the frequency $i$. The saturation coefficient $k_{sat,i}$ is equal to 1 for low-saturated circuits and $k_{sat,i} < 1$ for saturated circuits. The coefficient value for low saturated circuits is taken for the further calculations.

Electromotive differences and currents induced in stator windings induce electromotive difference and currents in rotor windings of the frequency $f_r = f_{grid}(N_1 - 1) = f_{grid}(N_2 - 1)$. The aforementioned electromotive differences are induced in the damper winding contour. It is important to mention that synchronous machine operation mode for high harmonics is asynchronous. The fields induced by high harmonic currents ($I_{N1}$, $I_{N2}$, $I_{N3}$) in stator and rotor windings create mutual induction field in the air gap of the same frequency spectrum and same space distribution flux wave. In accordance with Ampère's circuital law, magneto motive force can be described as (Eq. 4.16):

$$F_{res} = F_{N_i}^{rotor} + F_{N_i}^{stator}, \qquad (4.16)$$

where $F_{N_i}^{rotor} = |F_{N_i}^{rotor}|e^{j\psi_{N_i}^{rotor}}$ – rotor magnetomotive force; $F_{N_i}^{stator} = |F_{N_i}^{stator}|e^{j\psi_{N_i}^{stator}}$ – stator magnetomotive force; $\psi^{rotor}$ and $\psi^{stator}$ – rotor and stator phase angels. The mutual position of two vectors $F_{N_i}^{rotor}$ and $F_{N_i}^{stator}$ in space is characterized with the angle $\Delta\phi_{N_i}^{rotor} = \pi - \Delta\psi_{N_i}^{rotor}$. Thereby, Ampère's circuital law can be rewritten in the following way (Eq. 4.17):

$$|F_{N_i}^{res}| = |F_{N_i}^{rotor} + F_{N_i}^{stator}e^{j\Delta\phi_{N_i}^{rotor}}| \qquad (4.17)$$

The equation (4.17) can be used for estimating the screening effect of the fields induced by rotor currents at the frequency $N_i \approx N_i + 1$ on the fields induced by stator currents at the frequency $N_i$.

In the case of first harmonic prevalence in the frequency spectrum, the resultant magneto motive force in air gap is determined by voltage level at the terminals of the stator windings. In other words, $F_{N=1}^{stator}$ is a sum of complexes $F_{N=1}^{res}$ and $F_{N=1}^{rotor}$. The situation is different for analysis of high harmonic impact. Magneto motive force of stator windings $F_{N_i}^{stator}$ at the frequency $N_i$ is determined consistent with the current harmonic amplitude at the frequency $N_i$. The shielding efficiency is determined by the difference of rotor magneto motive force at the frequency $N_i$ and stator magneto motive force at the frequency $N_i$. It can be mathematically represented as in Eq. 4.18:

$$s_i = \frac{F_{N_i}^{res}}{F_{N_i}^{stator}} \tag{4.18}$$

The value $s_i$ is not constant and it differs with the harmonic growth by the following rule $0 < s_i < s1$, where $s_1$ is the screening efficiency at the main frequency. The value of $s_i$ can be found from Eq. 4.19:

$$s_i = \left( \frac{1 + X_{m,i}}{Z'_{2,i}} \right)^{-1}, \tag{4.19}$$

where $X_{m,i}$ – inductive resistance of magnetization circuit; $Z'_{2,i}$ – full impedance of the rotor windings correlated to stator windings.

Normally, the electric machine parameters are chosen in the way to achieve the minimum possible value of $s_i$ at typical non-compensated high harmonics, i.e. for fifth and seventh harmonics. In turn, it results in minimum additional losses in the active steel of the stator and rotor windings. High current harmonics caused by mutual inductance saturate the electric machine core. Nevertheless, reduction of $P_{adm}$ for the first harmonic prevents saturation.

The admissible power $P_{adm}$ can be found as the product of $P_{adm,st}$ (Eq. 4.14) and $P_{adm,steel}$ (Eq. 4.15) as shown in Eq. 4.20:

$$P_{adm} = P_{adm,st} P_{adm,steel} \tag{4.20}$$

It is important to note that $s_i$ value is inversely proportional to direct-axis armature reaction reactance $x_{ad}$ and is less than 0.1 (3). Thereby, it is enough to consider only $P_{adm,st}$ for achieving trustful results. If THD is less than 0.8, value of $P_{adm,steel}$ is equal to 0.95–0.9.

The equation (4.20) does not include losses in the rotor windings. Actual legal norms do not ask to perform the verification of rotor thermal conditions and only state that the temperature should not exceed 200–250 °C. The operational experience of modern powerful synchronous machines shows that the analysis of rotor windings overheating is required if $cos\phi < 0.7$.

The analysis of operating conditions for turbogenerator TBB-200 are represented in Table 4.7. The results for first five harmonics in case of single-phase transformer saturation are given.

The analysis shows that the synchronous machine admissible power $P_{adm}$ has to be reduced by ca. 50% in case of single-phase transformer saturation and by 25% in case of three-phase five-limb power transformer saturation. A group of three single-phase transformers are normally used as the step-up transformers at the large power plants. The generator and consequently connected step-up transformer are placed under a common circuit breaker on the high voltage side in the power distribution schemes for nuclear power plants and large thermal power plants. The placement of two blocks "generator – step-up transformer" under one circuit breaker is sometimes used at the voltage levels 330 kV and higher for reducing the total cost. It is also

**Table 4.7**

**The value of the main and additional ohmic losses in the stator windings (losses in the Cu)**

| Main losses, VAR | Additional losses, VAR | | |
|---|---|---|---|
| | $\Delta K_f = 0.075$ | $\Delta K_f = 0.15$ | $\Delta K_f = 0.25$ |
| 262,000 | 19,650 | 39,300 | 65,500 |
| 640,000 | 48,000 | 96,000 | 160,000 |
| 289,940 | 21,745 | 43,491 | 72,485 |
| 42,000 | 3,150 | 6,300 | 10,500 |
| 16,436 | 1,232 | 2,465 | 4,109 |
| 16,436 | 1,232 | 2,465 | 4,109 |

common to use enlarged blocks while hydro power plant design as they are often constructed in the areas with complex topography.

### 4.1.3 MEASUREMENT TRANSFORMERS – RELAY PROTECTION SYSTEM

Measurement (instrument) transformers are used to decrease the current and voltage values corresponded to HV and UHV networks up to the values suitable for measurement equipment by the means of galvanic disconnection. Both voltage (VT) and current (CT) measurement transformers exist. While the application reasons are the same for both types, the operational principles are different, and so are their vulnerability levels against GIC.

CT is characterized by transformation coefficient. The ratio between primary and secondary windings currents is determined by Eq. 4.21.

$$k_I = \frac{I_{PRI}}{I_{SEC}}, \tag{4.21}$$

The CT equivalent circuit is shown in Fig. 4.10.

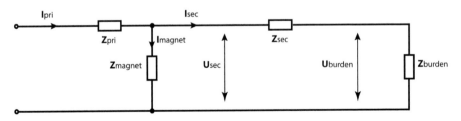

Figure 4.10: Current transformer equivalent circuit

The ratio of Eq. 4.21 is not constant and depends on the CT operation mode. The transformation from primary to secondary current is only linear in the certain part of magnetizing characteristic. When CT is driven to saturation, there will be almost no current on the secondary burden. Thereby, core's saturation prevents the transformer from accurately representing secondary current. The character of magnetizing characteristic is predetermined by transformers application purpose: instrument transformers that are used for relay protection devices and metering transformers those that determine the electricity price. Around 80% of installed capacity corresponds to the first type.

The application purpose specifies the maximum possible error: 2% in the second case and 5–10% in the first case. The Std. C37.110-2007 specifies the working area on magnetization curve, which ensures CT's accurate operation [44]. It states that the ratio between working voltage ($V_e$) and knee-point voltage ($V_k$) is less than 0.5 (Eq. 4.22). The knee-point voltage is the highest exciting voltage at which CT will still not be saturated.

$$\frac{V_e}{V_k} \leq 0.5, \tag{4.22}$$

The CT saturation depends on: the physical design; the amount of steel in the core; the connected burden; the winding resistance; the remnant flux; the fault level; the power system $X/R$ ratio [45]. The GIC leads the measurement transformer to saturation in the same way as the power transformer. The difference is in the entry point. In case of CT, the GIC as DC bias is coupled via the transmission line that is subjected to the ultra-low frequency magnetic field, while the GIC enters power transformers through the grounded neutral of a wye-connection. The core of measurement transformers for relay protection purposes is constructed from Si-Fe material and metering transformers core from Ni-Fe material. The test results showed that Ni-Fe alloy is more prone to saturation.

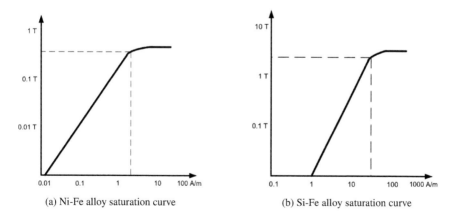

(a) Ni-Fe alloy saturation curve       (b) Si-Fe alloy saturation curve

Figure 4.11: Ni-Fe and Si-Fe alloys saturation curves

As it is depicted in Figs. 4.11a–4.11b, the Si-Fe alloy saturates at 10 times higher

DC current equal to 20 A. The CT under saturation poses threat to normal operation of relay protection system. The time taken to reach saturation in a CT can be substantially reduced if the transient DC offset of the fault current is of the same polarity as the DC bias current [44]. In case DC bias forces a partial saturation of CT, the time-difference method may fail to recognize CT saturation [46]. The secondary current, which is used an initial signal for relay protection, has a reduced amplitude under saturation. It may lead to relay protection misoperation. A common GMD effect on power grids was associated with protective relays misoperation that erroneously disconnected power system equipment responsible for power system state balance. Protective relay is a device that disconnects any element of a power grid experiencing a fault or abnormal behavior. There are two ways in which a protective relay can malfunction: fail to operate when expected or to operate when not expected to. Overcurrent, distance and current differential relays may be prone to GMD effects.

If currents amplitude is smaller than relay's pick-up value, it prevents relay operation. The time coordination of overcurrent relay is used in the way that the relay closest to the fault (with the highest current) will operate fastest. Hence, the fault will be cleared with the minimum amount of power grid isolated. The overcurrent relay logic scheme is presented in Fig. 4.12a. The relays time coordination is done with tendency to security. It takes into account circuit breaker opening time, relay over travel in electromechanical relays, and a safety factor to accommodate measurement errors [43]. CT saturation may cause failures with time current coordination. Large errors may result in loss of coordination or extreme cases in failure to trip. Time miss-coordination leads to network element overheating. Additionally, it results in not optimum fault clearance. Analysis showed that the operation differs for different types [47].

Distance protection may experience overreach and under-reach problems due to inaccurate current phase measurements. Since distance impedance is calculated by relay using the amplitude of current and voltage samples, the calculated impedance during saturation period will be modified. The logic scheme is presented in Fig. 4.12b. The two critical cases are the faults close to the relay location and the fault close to zone 1. Faults close to zone 1 are much more critical. GIC as DC component may cause underreaching [48].

The most serious consequences of CT saturation lay in the field of current differential protection operation since its logic is based on Kirchoff's law (Fig. 4.12c). Most differential protection relays are based on a percentage restraint characteristic to restrain tripping for high through currents (Fig. 4.13). The pick-up value is chosen with respect to accepted error and is expressed as the restraint current $I_{trip} = I_{diff_0} + k_{restrain} \left( I_{restrain} - I_{restrain_0} \right)$. Minor CT's saturation can be compensated by percentage restraint. Heavy saturation leads to miss operation. In addition, high harmonics disable the fast operation of a differential relay. The relays default on their high-current unrestrained element alone. Power transformers core materials highly impact the magnitude of the second harmonic [49]. However, the GIC does not pose any new risks to the operating performance of converter transformer differential protection [50].

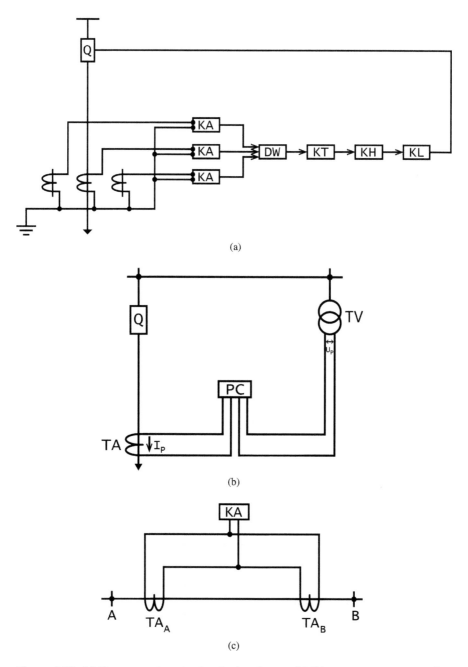

Figure 4.12: (a) Overcurrent protection logic scheme; (b) Distance current protection logic scheme; (c) Differential current protection logic scheme

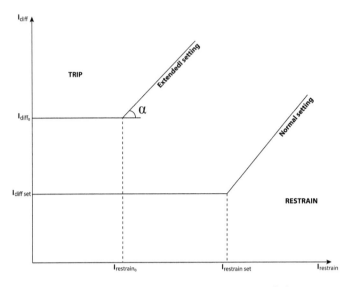

Figure 4.13: Percentage restraint characteristic

As well as for the CT, VT transformation error is determined by core material magnetic permeability, core's construction and secondary load value. In the same manner, voltage transformer accuracy depends on the magnetization current. Regardless of application purpose, the whole range of VTs is designed from Si-Fe alloy, which is more resilient to saturation.

Possible instrumental transformer loss is less critical than power transformer loss in terms of restoration cost and period. The power transformer unit can be replaced in the period of 12–18 months. The replacement cost can reach USD 20 million. Contrary, the instrumental transformer replacement cost is in the range of several tens of thousands of US dollars. Also, the restoration period is relatively short. Nevertheless, instrumental transformer outage can trigger sufficient economic loss. Since power system equipment is protected by the means of relay protection system, signal accuracy is of high importance. In their turn, relay protection devices, receive data about power system state from instrumental transformers, which have increased error in saturation mode. Thereby, the input data can represent power system state inaccurately. Thus, power system controllability can be deceased. A special attention should also be paid to modern digital relays in order to prevent the 2003 situation in Norway [51].

### 4.1.4 GEOMAGNETICALLY INDUCED CURRENT IMPACT ON OTHER POWER SYSTEM EQUIPMENT

Other power system equipment with magnetic core may also be influenced by GIC flow over its windings. The detailed analysis of GIC impact on shunt reactors operation is represented in [52]. Shunt reactor's sensitivity to GIC effects is rather low.

Due to its construction features, a relatively high GIC is required to cause shunt rector saturation. The current level is far beyond the maximum specified level. In power transformers, gaps have to be minimized in order to ensure efficient voltage and power transformation between the windings. Contrary to power transformers, shunt reactors contain wider non-magnetic gaps between core steel packets. Circuit breakers also contain magnetic core, but air gaps between core-steel packets are wider than in shunt reactors. Thereby, it is assumed that their susceptibility to GMD is admissible.

Modern power grids are characterized by high penetration of high voltage direct current (HVDC) convertors. HVDC convertors bring no barrier to GIC flow over its elements. The GIC flow by itself has no negative impact on HVDC convertor operation, since possible GIC amplitudes are negligibly small compared to nominal currents (kAs). The danger to HVDC converter might be posed from the side of abnormal high harmonics distribution due to power transformer saturation. The HVDC converters by themselves generate the current harmonics on the AC-side. Therefore, the means for limiting the current distortion to an acceptable level are already predesigned.

The orders of the generated harmonics are determined by the pulse number of the converter configuration defined as the number of simultaneous commutations per cycle of the nominal frequency. The modern HVDC systems consist of a 12-pulse converter, formed by connecting two 6-pulse bridges. One group has a *wye − wye* connected converter transformer, and the other group has a *wye − Δ* connected converter transformer. It is convenient to classify the HVDC converter harmonics as characteristic and non-characteristic harmonics. The characteristic harmonics are those with the order $pk \pm 1$, where $k$ is any positive integer; $p$ is a pulse number. The current on the secondary side $i_2$ can represented as the function of firing angle $\alpha$ and commutation angle $\gamma$ (Eq. 4.23):

$$i_2(\alpha, \gamma) = \frac{\sqrt{3}E_{2m}}{2X_\gamma}(cos\alpha - cos(\alpha + \gamma)) \quad (4.23)$$

The amplitude of the $n$th harmonic is given by Eq. 4.24:

$$I_n = \frac{\sqrt{6}I_d}{n\pi} \quad (4.24)$$

The harmonics caused by imperfect system conditions are called non-characteristic. In other words, the static converters generate harmonic orders and magnitudes not predicted by the Fourier series of the idealized waveforms. The main causes are listed in [53]. The compensation of non-characteristic harmonics is made by installing tuned filters. The single tuned filter is a serial RLC circuit that is tuned at one harmonic frequency. Its impedance is described in Eq. 4.25:

$$Z_f = R + j\left(\omega L - \frac{1}{\omega C}\right) \quad (4.25)$$

The series capacitors are also affected by high harmonic distortion. They perform as low-impedance paths for harmonics. Series capacitor's impedance decreases with the frequency growth leads to the growth of currents amplitude. This can result in over-voltages. Series capacitors connected using the scheme "star with isolated neutral" have higher robustness to GMDs, since harmonics with the order $3k$ are blocked.

The transmission line operation conditions are predominantly limited by the admissible thermal forces which are chosen in accordance with its sag. The performed investigation showed that a GIC equal to 2 kA or higher can lead to excessing admissible values for transmission line sag. This current level is also far beyond the maximum specified level. The other problem is the non-uniform distribution of current into the conductor section. In other words, current will tend to flow closer to the surface of the conductor, which leads to uneven heat distribution in the conductor. The power of the flux of electromagnetic energy penetrating into the conductor through its surface and emerging in the conductor in the form of heat $P$ can be calculated as in Eq. 4.26:

$$P = \frac{l}{u}\sqrt{\frac{\omega\mu}{2\gamma}}I^2, \tag{4.26}$$

where $l$ is the conductor's surface through which electromagnetic wave penetrates; u is the perimeter of the conductor's section; $\omega$ is the current's frequency; $\mu$ is the magnetic permeability; and $\gamma$ is the propagation coefficient.

The relative comparison of power system equipment susceptibility to direct and indirect GMD effects is presented in Table 4.8. The robustness of each equipment type to the GIC impact is represented in the first column. The level of impact on system operation in case of unit loss is shown in the second column and repair cost is given in the third column. Repair cost includes also replacement cost in case an equipment unit cannot be repaired swiftly.

**Table 4.8**

**Comparison of power system equipment susceptibility to geomagnetically induced currents**

| System equipment | Equipment robustness | System effect | Repair cost |
| --- | --- | --- | --- |
| Power transformer | Low | High | High |
| Instrument transformer | Medium | High | Low |
| Synchronous machines | Medium | High | High |
| Shunt reactors | High | High | High |
| Circuit breakers | High | Medium | High |
| Capacitors | High | High | High |
| DC substations | High | High | High |
| Transmission lines | High | High | Medium |

## 4.2 MODERN APPROACHES OF GEOMAGNETICALLY INDUCED CURRENTS INTEGRATION INTO THE POWER GRID STATE CALCULATION

Several approaches can be distinguished in power grid modeling. The most straightforward method is knowledge-based investigations, in which data from the past events analyzed by experts is used to improve understanding of system vulnerability and risk. It is a purely data-driven method. The other group of methods is so-called "best practices" which consists of (1) security-constrained assessment, (2) online risk-based assessment, (3) cascading outage assessment. Security-constrained assessment is made with respect to constrained reliability parameters and total cost optimization. Online risk-based assessment is used to assess whether the real-time state of the power system is secure with respect to several indicators, i.e. overloads, voltage instabilities, cascading overloads [54]. Cascading outage assessment model studies the propagation of a disturbance from a local incident to the system level. Based on the analysis results, power grid planning and operation is performed following the principle of minimum cost $C(x_i)$ (Eq. 4.27).

$$C(x_i) = \sum \left( C_t^c(x_i) + C_t^o(x_i) \right) \cdot (1+d)^{-t} \overrightarrow{x_i} min, \qquad (4.27)$$

where $C^c(x_i)$ is the capital cost for $x_i$ power grid; $C^o(x_i)$ is the operational cost for $x_i$ power grid; and $d$ is the discount rate.

The advanced approach for power grid planning is proposed in [55]. Two criteria are introduced: total cost minimization $C(x_i)$ and probability of power shortage minimization $J(x_i)$ (Eq. 4.28).

$$\begin{aligned} C(x_i) \overrightarrow{x_i} min \\ J(x_i) \overrightarrow{x_i} min \end{aligned} \qquad (4.28)$$

GICs flow over network elements may lead to detrimental effects on the grid's stability. Pathways to a failure are depicted in Fig. 4.3. Power transformer half-cycle saturation presents a challenge for power system control by unit heating, harmonics distortion, and reactive power deficit. Industry stakeholders agree that reactive power consumption is the most concerning physical effect of space weather on the power grid [56]. The physics of reactive power consumption growth is the following. Large magnetization current under half-cycle saturation lags the voltage of the system by $90°$, thus leading to an inductive circuit which consumes reactive power and creates a positive sequence reactive power requirement from the system [57]. The reactive power deficit is a non-linear function of a GIC and a power transformer type. Based on the previous conclusions, it is possible to say that power transformer with a larger core is more robust. Reactive power deficit is not only the function of GIC amplitude, but also directly proportionally to the AC voltage [58]. It is recommended to distinguish fundamental and high harmonics [59]. Only fundamental frequency positive-sequence voltages and currents are considered for reactive power deficit calculation, since only fundamental lagging currents have significant impact on voltage stability. However, an approach to bridge the gap between GIC flow/loadflow and harmonic analysis is given in [60]. Since GIC is a quasi-DC, it has ultra low frequency change. These currents together with corresponding power flows distribute through geographically separated transformers and produce power swings. The GMD does not necessarily cause a power grid failure. It can impact power grid operation on a moderate way. For instance, the transmission losses and the incidence of constrained transformers in PJM's 500 kV transmission system were also statistically related with the GIC proxy [61].

The analysis of GIC impact on power system state consists of three tiers. The first tier is the evaluation of a GIC threshold value above which transformers may be at risk. The second tier is the corresponding power flow simulation with an increased reactive power flow and potentially disconnected power system equipment. The last tier is the analysis of high harmonics impact. The GIC impact on power flows in the grid was the focus of several research including [62] and [63], following different single element contingencies [64], and others. The first study on the incorporation of the GIC into power flow modeling is done in [34]. The nature of the relationship between the GIC and transformer VAR demand was studied using the EMTP by injecting the DC currents into the windings of saturable reactors under sinusoidal conditions. The relationship can be represented using the formula (4.29).

$$Q_{loss} = U_{kV} k I_{GIC}, \tag{4.29}$$

where $Q_{loss}$ is the transformer's GIC-related reactive power loss in MVAr, $U_{kV}$ is the terminal voltage in kV, $k$ is a transformer specific constant, and $I_{GIC}$ is an effective GIC introduced in Section 4.1.1.

Equation 4.29 can be expressed in per unit (pu), since the power gird calculation is often done in this way (Eq. 4.30):

$$Q_{loss} = U_{pu} K I_{GIC}, \tag{4.30}$$

where $U_{pu}$ is the terminal voltage per unit, and coefficient $K$ in MVAr/A is specified by the transformer.

It is proposed to take the value $K$ equal to 2.8, hence the value is a function of the saturation curve shape. Since it is difficult to obtain precise saturation curves for each power transformer, Eq. 4.30 can be modified by assuming nominal voltage in the definition of $K$. Then the constant value needs to be scaled based upon the transformer's actual maximum nominal kV level [65]. The reactive power loss equation then becomes (Eq. 4.31).

$$Q_{loss} = U_{pu} K \left( \frac{U_{nom,kV}}{U_{nom,kVassumed}} \right) I_{GIC}, \tag{4.31}$$

where $U_{nom,kV}$ is the nominal voltage of the highest transformer winding in kV, and $U_{nom,kVassumed}$ is the assumed nominal voltage. If $K$ for a specified transformer is known, then Eq. 4.30 is used, and assumed voltage value is equal to nominal voltage value.

The most common technique for solving non-linear power balance equations is Newton-Raphson solution. It uses the first order Taylor series to linearize the power balance equations. The linear system of equations can be expressed as in Eq. 4.32:

$$\begin{bmatrix} \Delta\Theta \\ \Delta|\mathbf{V}| \end{bmatrix} = -\mathbf{J}^{-1} \begin{bmatrix} \Delta\mathbf{P} \\ \Delta\mathbf{Q} \end{bmatrix} \tag{4.32}$$

where $\Theta$ is the voltage phase angles of the buses; $|\mathbf{V}|$ is the voltage magnitude; $\Delta\mathbf{P}$ is a vector containing all real power imbalances; $\Delta\mathbf{Q}$ is a vector containing all reactive power imbalances; and $\mathbf{J}$ is the square Jacobian matrix defined as in Eq. 4.33:

$$\mathbf{J} = \begin{bmatrix} \frac{\partial\Delta\mathbf{P}}{\partial\Delta\Theta} & \frac{\partial\Delta\mathbf{P}}{\partial\Delta|\mathbf{V}|} \\ \frac{\partial\Delta\mathbf{Q}}{\partial\Delta\Theta} & \frac{\partial\Delta\mathbf{Q}}{\partial\Delta|\mathbf{V}|} \end{bmatrix} \tag{4.33}$$

The reactive power deficit associated with GIC is considered in Jacobian matrix as in Eq. 4.34. The entries of Jacobian, which correspond to the partial derivative of the reactive power to voltage magnitudes of the same bus, are modified. The other entries of Jacobian remain unchanged.

$$\frac{\partial\Delta Q_i}{\partial\Delta|V_i|} \leftarrow \frac{\partial\Delta Q_i}{\partial\Delta|V_i|} - KI_i^{GIC} \tag{4.34}$$

Traditionally, the GIC impact on power grid state stability is considered as a small-signal stability problem. Small-signal stability studies a system response to small perturbations about a particular operating point. The slow GIC change is assumed in this case, and the critical GIC at which the power flow equations have no real solution is identified. Two aspects contradict this simplification. Electromagnetic processes in power system equipment are relatively fast. Fast rise times in transformer neutral currents have been observed in measurements [23]. Elevated geoelectric fields which correspond to high GICs persist for relatively short time over GMD duration. The electric fields sometimes have rise times of approximately 30 seconds [66]. Large disturbance (dynamic) voltage stability considers the time domain response of a system after a large disturbance such as a generator outage. The large disturbance in a GMD context is a time variation in the GIC [67]. The problems of dynamic power grid stability under GMD for a five-bus test scheme is discussed in [68].

The accurate estimate of GMD scenario obtained from situational awareness centers is required for proper state estimation (Section 5.2). In other case, the error may propagate to subsequent operations such as reactive power losses calculation, admissible power flow assessment, etc. The evaluation of an average absolute error defined as a difference between the estimated and actual voltage magnitudes is given in [69]. Both power grid parameters data and geophysical condition can cause a calculation error. Required information about power grid elements is beyond those for a typical AC power flow analysis. The US Department of Energy concluded that the absence of information on substation grounding and transformer configuration details is critical. Missing information on power transformer's design can be extracted from design review documentation for transformers younger than 20 years. In other cases, data can be extracted from test reports and outline drawings. In both cases, data can be purchased from the manufacturer. The average absolute error as a function of storm's magnitude grows with the amplitude increases from 0.004 pu for 1 V/km to 0.03 pu for 9 V/km. Since the transmission line susceptibility to geoelectric field depends on the mutual orientation of transmission line and the field, field direction is another source of a calculation error. Hence, the absolute error at a constant field amplitude varies periodically with varying storm direction.

The GIC as ultra-low frequency current is characterized by low information entropy. According to the mathematical theory of information, the amount of information per unit of time obtained from registered GIC is small. This complicates the GMD parametrization using modern instrumentation. The GIC is an example of a random process. Therefore, it can be analyzed using the standard mathematical apparatus of probability theory. Consequently, it can be well predicted at time intervals of several seconds or more. The deterministic component prevails, and the process itself can be considered quasi-determined. In compliance with the mathematical theory of information, it is a random component that carries information. However, its share is insignificant. Two scenario exist to overcome this problem: sufficient growth of the time spent for GMD parametrization, which limits the emergency response efficiency, or GMD parametrization using the signal with reduced accuracy, which leads to higher economic losses.

Two factors are important in determining the amount of information obtained from GIC registration: narrow band of the recorded signal and the signal-to-noise ratio in the measuring channel. These two factors are reflected in the formula for coefficient of variation (Eq. 4.35). It represents the maximum mutual entropy of the GIC and recorded signal. In other words, it represents the maximum amount of information that can be derived from GIC for GMD parametrization.

$$CV = \int_{-\infty}^{\infty} log_2\left(\eta^2\left(\omega\right)+1\right)d\omega \qquad (4.35)$$

where $\eta\left(\omega\right)$ is the signal-to-noise ratio in the measuring channel.

Eq. 4.35 is accurate, though the improper integral calculation involves computational difficulties. The approximation of formula (4.35) is proposed in Eq. 4.36, for which this drawback is absent. The approximation error can be neglected due to its small value in case $N > 1,000$.

$$CV = \frac{2\Delta F}{N}\sum_{k=0}^{N-1} log_2\left(\frac{|U\left(k\right)|^2}{\gamma}+1\right) \qquad (4.36)$$

where $U\left(k\right)$ is the discrete Fourier transform of the recorded signal samples; $\Delta F$ is the GIC frequency band; and $N$ is the window size for discrete Fourier transform of the recorded signal.

In the majority of measurement systems, signal is much stronger than the noise level. However, the GIC frequency lies in the same range as thermal noise of the modern measurement equipment. The value $\eta\left(\omega\right)$ for all $\omega$ is small, which results in small coefficient of variation values. It is possible to consider the registered signal as low-entropic. In fact, the task of GMD parametrization is reduced to the well-known electrical engineering problem of the constant component measurement, which currently does not have a satisfactory solution. It is driven by the fact of a sharp increase of the thermal noise spectral density at a frequency tending to zero.

It can be summarized that GIC is a separate class of signals. New methods are required for their processing that allow us to separate time dependencies having the similar spectral characteristics. Due to small coefficient of variation values, the task of GMD parametrization looks sophisticated, since GICs do not provide sufficient amount of information to measurement equipment. Researches in two directions are suggested to solve this problem. Firstly, the economic loss costs and mitigation action costs should be optimized. Secondly, it is essential to optimize the measurement equipment by taking into account the low-entropy features of the GIC.

Measures aiming at noise cancellation are characterized by high cost and low efficiency. The compensation method is preferable. It is important to note that GIC and recorded signal have different time dependencies. The GIC does not contain any noise. Noise is added to the signal during the registration process. Therefore, it can be compensated. Since GIC has low entropy, it is possible to predict it with high accuracy at small intervals as well as thermal noise.

The scheme of a proposed compensation device is depicted in Fig. 4.14. The GIC signal is alternately fed to analog-to-digital converter (ADC) 1 or 2, where it is sampled and quantized. As any measurement device, ADC is susceptible to external (temperature) and internal (aging) impacts, which influence gain signal and zero level. The influence on other parameters is not that significant. Despite the fact that registered processes have a very narrow spectrum ($\ll 1$ Hz), the sampling frequency is chosen high enough (e.g. 1 kHz). Due to the fact of signal's low-entropy, it is possible to switch the inputs between ADC1 and ADC2 every few minutes. In this case, one of the ADC performs signal calibration and the other one digitalizes it. Whilst calibrating the signal, ADC is first switched to the ground, which is the zero level standard $U_c$, and to the reference voltage $U_e$. Whenever the signal is fed to the calibrated ADC, the compensation algorithm is processed. This allows removing the spurious DC component from GIC signal. In other words, it compensates the temperature gain coefficient $k$.

The signal after analog-to-digital conversion can be mathematically represented as in Eq. 4.37.

$$\tilde{u}(t) = k(u(t) + U_c), \qquad (4.37)$$

where $\tilde{u}$ is the distorted signal due to temperature drift; and $u(t)$ is the true GIC signal.

If ADC is grounded, it captures the signal $U_0 = kU_c$, and signal is described as $U_1 = k(U_e + U_c)$, when ADC is switched to the reference voltage (Eq. 4.38).

$$\begin{cases} U_0 = kU_c \\ U_1 = k(U_e + U_c) \end{cases} \qquad (4.38)$$

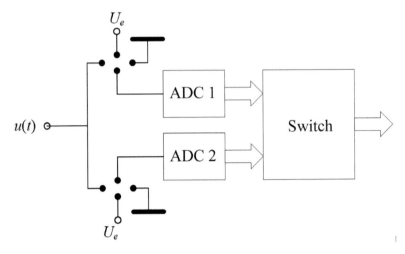

Figure 4.14: Compensation device scheme

The values of the spurious constant component and the current gain signal can be described as in Eqs. 4.39.

$$\begin{cases} U_c = \frac{U_e U_0}{U_1 - U_0} \\ k = \frac{U_1 - U_0}{U_e} \end{cases} \tag{4.39}$$

Further, the true GIC signal $u(t)$ can be found as Eq. 4.40:

$$u(t) = \frac{\tilde{u}(t)}{k} - U_c \tag{4.40}$$

The described technique can be improved by using digital filtering as well as various algorithms for edge distortions averaging and compensation. In addition, it is possible to create a scheme using only one ADC when the missing signal is reconstructed using interpolation methods.

Power flow analysis of real power grids under GIC is the focus of ongoing research in the regions considered to be prone to GMD. For instance, the stability of American Electric Power company, which is one of the largest electric utilities in US, for certain GMD scenario was analyzed in [70]. The analysis included GMD scenario definition, GICs calculation followed by a consequent increase of reactive power losses and their integration into the power flow model. The impact of high harmonics distortion was not performed. The output of the research was the determination of the power transformers at risk based on the high GIC value distribution. It was defined that the lowest electric field that may cause voltage collapse is 13 V/km. In contrast, analysis of a simplified Hydro-Québec grid model showed that voltage avalanche can be reached at the geoelectric field of 2 V/km [71]. The evolution of the Finnish grid showed that there is a need for 660 MVAr reactive power capacity to support the annual flow of 1200 A altogether via the transformer neutrals to and from the earth. This means from the system point of view that a voltage collapse and system black-out could occur roughly once in 20 years if an extreme case takes place [51]. The simulation for AltaLink network, Canada, revealed power transformers at risk [72].

Transformation of traditional power grids poses new challenges to their operation under GMDs. One of the significant features of modern power grids is the prevalence of decentralized renewable generation sources. Distributed generation has long been technically possible. However, we are nearing a tipping point, beyond which, for many applications, distributed generation will be the least costly way to provide electricity [73]. By 2015, wind and solar generation represent roughly 10% of the world's installed capacity. It was shown that a moderate neutral GIC of 37 A in main output transformer of wind power plant can cause a harmonic distortion beyond the standard limit of 5% [74].

## 4.3 CONCLUSION

The GMD differs from other natural hazards as it does not have a unique way to endanger power grid operation. For instance, power grid failure caused by earthquakes is caused by strong ground motion which results in foundations of buildings and transmission towers failures. Risk associated with floods is a reduced shear strength at the landslides sites. Normally, only few vulnerable nodes are responsible for power grid's blackout. Including the GMD to the list of threats allows us to make the comprehensive list of contingencies. The overall process of power grid state assessment to the GIC risk is presented in Fig. 4.15.

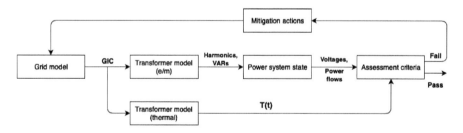

Figure 4.15: Algorithm for GIC impact assessment on power grid state

Depending on the GMD type, the same power grid equipment can be vulnerable in different ways. This is a multi-criteria problem. The detailed overview of other critical factors of different nature will be given in Section 6.3. In terms of power grid equipment robustness to GMDs, the GIC value and geomagnetic field rate-of-speed can be a measure to evaluate the possible threat scenario. GMDs can lead either to system-wide loss or postponed outage. Three following scenarios exist:

- an intense GIC that flows for a short duration can lead to cascade failures as during Hydro-Québec blackout

- long-lasting high GIC can cause severe transformer overheating, leading to failure in days (as it was the case for the loss of power transformer at Salem nuclear power plant)
- moderate GIC can initiate localized degradation of power transformer isolation that continues even after GIC flow ends (as it was the case for the power transformer loss in South Africa)

The carried analysis helped to elucidate important aspects of the power grid's vulnerability to the GMDs. Damage type is correlated to the damage to transmission and generation equipment from GICs and/or relay protection system. Contributing factors include equipment type, its construction scheme and isolation characteristics, on one side, and the GIC parameters, on another side. Most vulnerable equipment to quasi-DC flow is power transformers. Power equipment protected from DC excitation may trip in the time of an event. Recovery time is driven by the territory affected by the blackout or by delayed effects. Power to areas serviced by the equipment which has only tripped offline could be restored within less than 24 hours after the end of the storm. Repairs of damaged equipment may take several months.

## REFERENCES

1. Karagiannis, G. M., Chondrogiannis, S., Krausmann, E., Turksezer, Z. I. (2017). Power grid recovery after natural hazard impact. *Joint Research Center: European Union.*
2. Department of Energy (2014). *Large power transformers and the U.S. Electric grid.* Washington D.C., Department of Energy.
3. Royal Academy of Engineering (2013). *Extreme space weather impact on engineered systems and infrastructure.* London.
4. Hapgood, M., Thomson, A. (2010). Space weather: Its impact on Earth and implications for business. *Lloyd's 360 Risk Insight.* London.
5. Kirkham, H., Makarov, Y. V., Dagle, J. E., DeSteese, J. G., Elizondo, M. A., Diao, R. (2011). *Geomagnetic storms and long-term impacts on power systems* (No. PNNL–21033). Pacific Northwest National Lab. (PNNL), Richland, WA (United States).
6. Kulkarni, S. V., Khaparde, S. A. (2017). *Transformer Engineering: Design, Technology, and Diagnostics.* CRC press.
7. Li, X., Wen, X., Markham, P. N., Liu, Y. (2010). Analysis of nonlinear characteristics for a three-phase, five-limb transformer under DC bias. *IEEE Transactions on Power Delivery, 25(4),* 2504–2510.
8. Dasgupta, I. (2009). *Power Transformers Quality Assurance.* New Age International.
9. Chisepo, H. K., Gaunt, C. T., Borrill, L. D. (2019). Measurement and FEM analysis of DC/GIC effects on transformer magnetization parameters. In *2019 IEEE Milan PowerTech*, 1–6. IEEE. doi: 10.1109/PTC.2019.8810423
10. Harrison, C. W., Anderson, P. I. (2016). Characterization of grain-oriented electrical steels under high DC biased conditions. *IEEE Transactions on Magnetics, 52(5),* 1–4.
11. Chisepo, H. K., Gaunt, C. T., Borrill, L. D. (2019). Measurement and FEM analysis of DC/GIC effects on transformer magnetization parameters. In *2019 IEEE Milan PowerTech*, 1–6. IEEE.

12. Girgis, R. S., Ko, C. D. (1992). Calculation techniques and results of effects of GIC currents as applied to large power transformers. *IEEE Transactions on Power Delivery, 7(2)*, 699–705.

13. Girgis, R., Vedante, K. (2013). Methodology for evaluating the impact of GIC and GIC capability of power transformer designs. In *2013 IEEE Power & Energy Society General Meeting*, 1–5. IEEE.

14. Rezaei-Zare, A. (2013). Behavior of single-phase transformers under geomagnetically induced current conditions. *IEEE Transactions on Power Delivery, 29(2)*, 916–925.

15. Mousavi, S. A., Bonmann, D. (2017). Analysis of asymmetric magnetization current and reactive power demand of power transformers due to GIC. *Procedia Engineering, 202*, 264–272.

16. Chiesa, N., Lotfi, A., Hoidalen, H. K., Mork, B., Rui, O., Ohnstad, T. (2013). Five-leg transformer model for GIC studies. In *International Conference on Power Systems Transients (IPST2013) in Vancouver, Canada*.

17. Cherry, E. C. (1949). The duality between interlinked electric and magnetic circuits and the formation of transformer equivalent circuits. In *Proceedings of the Physical Society. Section B, 62(2)*, 101.

18. Sokolova, O., Burgherr, P., Collenberg, W. (2015). Solar storm impact on critical infrastructure. *Safety and Reliability: Methodology and Applications*, 1515–1521. doi: 10.1201/b17399-207.

19. Mousavi, S. (2015). *Electromagnetic modelling of power transformers for study and mitigation of effects of GICs*. Doctoral dissertation, KTH Royal Institute of Technology, Stockholm.

20. McNutt, W. J. (1990). The effect of GIC on power transformers. *IEEE PES Summer meeting July 17, 1990*, 17:32–37.

21. Kappenman, J.G. (1996). Geomagnetic storms and their impact on power systems. *IEEE Power Engineering Review*, 16:5.

22. IEC (2017). *Standard IEC60034-1 Roating electrical machines - Part 1: Rating and performance*. International Electrotechnical Commission, Geneva.

23. Kappenman, J. (2010). *Geomagnetic Storms and Their Impacts on the US Power Grid*. Goleta, CA, Metatech.

24. Girgis, R., Vedante, K. (2012). Effects of GIC on power transformers and power systems. In *PES T&D 2012*, 1–8. IEEE.

25. Samuelsson, O. (2013). *Geomagnetic disturbances and their impact on power systems*. Division of Industrial Electrical Engineering and Automation Lund University, Sweden.

26. Gaunt, C.T., Coetzee, G. (2007). Transformer failures in regions incorrectly considered to have low GIC-risk. In *2007 IEEE Lausanne Power Tech*, 807–812. IEEE.

27. Béland, J., Small, K. (2004). Space Weather Effects on Power Transmission Systems: The Cases of Hydro-Québec and Transpower New Zealand Ltd. In *Effects of Space Weather on Technology Infrastructure*, 287–299. Springer, Dordrecht

28. Cigre Working Group 12.18 (2002). *Guidelines for life management techniques for power transformers*. Cigre.

29. Board, I. (1995). IEEE guide for loading mineral-oilimmersed transformers. *IEEE Std C, 57*, 1–112.

30. Oommen, T. V. (1981). Cellulose insulation materials evaluated by degree of polymerization measurements. In *Proc. 15th Elect./Electron. Insul. Conf., 1981*, 257–261.

31. Moodley, N., Gaunt, C. T. (2017). Low energy degradation triangle for power transformer health assessment. *IEEE Transactions on Dielectrics and Electrical Insulation*,

*24(1)*, 639–646. doi: 10.1109/TDEL.2016.006042.

32. Moodley, N., Gaunt, C. T. (2012). Developing a power transformer low energy degradation assessment triangle. In *IEEE Power and Energy Society Conference and Exposition in Africa: Intelligent Grid Integration of Renewable Energy Resources (PowerAfrica)*, 1–6. IEEE.

33. Lahtinen, M., Elovaara, J. (2002). GIC occurrences and GIC test for 400 kV system transformer. *IEEE Transactions on Power Delivery, 17(2)*, 555–561.

34. Albertson, V. D., Kappenman, J. G., Mohan, N., Skarbakka, G. A. (1981). Load-flow studies in the presence of geomagnetically-induced currents. *IEEE transactions on power apparatus and systems, (2)*, 594–607.

35. Zheng, K., Boteler, D., Pirjola, R. J., Liu, L. G., Becker, R., Marti, L., Guillon, S. (2013). Effects of system characteristics on geomagnetically induced currents. *IEEE Transactions on Power Delivery, 29(2)*, 890–898.

36. Patil, K. (2014). Modeling and evaluation of geomagnetic storms in electric power system. Paper C4-306 presented at *CIGRE 2014, Paris*.

37. Girgis, R., Vedante, K., Burden, G. (2014). A process for evaluating the degree of susceptibility of a fleet of power transformers to effects of GIC. In *2014 IEEE PES T&D Conference and Exposition*, 1–5. IEEE.

38. Elovaara, J., Lindblad, P., Viljanen, A., et al. (1992). Geomagnetically induced currents in the Nordic power system and their effects on equipment, control, protection and operation. *CIGRE Colloq., Aug. 30 – Sep. 1992, Paris, France.*

39. Molinski, T. S., Feero, W. E., Damsky, B. L. (2000). Shielding grids from solar storms [power system protection]. *IEEE Spectrum, 37(11)*, 55–60.

40. Pirjola, R., Kauristie, K., Lappalainen, H., et al. (2005). Space weather risk. *Space Weather 3*, 1–11.

41. Arsenjev, I.A., Boguslawsky, I.Z., Popov, V.V., et al. (2013). Features of creation and operation of powerful alternating current machines in off-line power supply networks. *St. Petersburg State Polytechnical University Journal 183*, 41–48. [in Russian]

42. Boguslawsky, I., Korovkin, N., Hayakawa, M. (2017). *Large AC Machines*. Tokyo, Springer.

43. Smith, T., Richard, H. (2013). Current transformer saturation effects on coordinating time interval. *IEEE Transactions on Industry Applications 49*, 825–831.

44. IEEE (2007). *IEEE Guide for the application of current transformers used for protective relaying purposes, IEEE Std C37.110-2007*. IEEE Power Engineering Society, USA.

45. Ozgonenel, O. (2013). Correction of saturated current from measurement current transformer. *IET Electric Power Applications, 7(7)*, 580–585. doi: 10.1049/iet-epa.2013.0105.

46. Lu, G., Huang, T., Zhang, F., Zheng, T., Liu, L. (2016). The effects of the current transformer saturation on mal-operation under the DC magnetic bias caused by HVDC. In *12th IET International Conference on AC and DC Power Transmission (ACDC 2016)*, 1–6.

47. Mattei, A.K., Grady, W.M. (2019). Response of Power System Protective Relays to Solar and HEMP MHD-E3 GIC. In *2019 72nd Conference for Protective Relay Engineers (CPRE)*, 1–7.

48. Holbach, J. (2006). Modern Solutions to Stabilize Numerical Differential Relays for Current Transformer Saturation during External Faults. In *2006 Power Systems Conference: Advanced Metering, Protection, Control, Communication, and Distributed Resources*, 257–265. IEEE. doi: 10.1109/PSAMP:2006.285398.

49. Mekic, F., Girgis, R., Gajic, Z., et al. (2007). Power transformer characteristics and their effects on protective relays. In *2007 60th Annual Conference for Protective Relay Engineers*, 455–466.

50. Zhao, Y., Crossley, P. (2020). Impact of DC bias on differential protection of converter transformers. *International Journal of Electrical Power & Energy Systems, 115*. doi: 10.1016/j.iiepes.2019.105426

51. Elovaara, J. (2007). Finnish experiences with grid effects of GIC's. *Space Weather*, 311–326. Springer, Dordrecht.

52. Ngnegueu, T., Marketos, M., Devaux, F., et al. (2012). Behavior of transformers under DC/GIC excitation: Phenomenon, impact on design/design evaluation process and modeling aspects in support of design. In *Proceedings of the CIGRE, 2012, Paris, France*.

53. Arrillaga, J., Liu, Y.H., Watson, N.R. (2007). *Flexible power transmission: the HVDC options*. John Wiley & Sons.

54. Ni, M., McCalley, J. D., Vittal, V., Tayyib, T. (2003). Online risk-based security assessment. *IEEE Transactions on Power Systems 18*, 258–265.

55. Belyaev, N., Egorov, A., Korovkin, N., Chudny, V. (2018). Economic aspects of ensuring the capacity adequacy of electric power systems. In *E3S Web of Conferences 58*, 01010. EDP Sciences. doi: 10.1051/e3sconf/20185801010.

56. Worman, S., Taylor, S., Onsager, T., Adkins, J., Baker, D. N., Forbes, K. F. (2018). The Social and Economic Impacts of Moderate and Severe Space Weather. In *Extreme Events in Geospace*, 701–710. Elsevier.

57. Mkhonta, S., Murwira, T. T., Oyedokun, D. T. O., Folly, K. A., Gaunt, C. T. (2018). Investigation of Transformer Reactive Power and Temperature Increases Under DC. In *2018 IEEE PES/IAS PowerAfrica*, 595–600. IEEE. doi: 10.1109/PowerAfrica.2018.8520998.

58. Walling, R.A., Khan, A.N. (1991). Characteristics of transformer exciting-current during geomagnetic disturbances. *IEEE Transactions on Power Delivery 6*, 1707–1714.

59. Emanuel, A.E. (1990). Powers in nonsinusoidal situations–a review of definitions and physical meaning. *IEEE Transactions on Power Delivery 5*, 1377–1389.

60. Adhikari, S., Mueller, D., Walling, R., O'Laughlin, A. J. (2017). A comprehensive study of geomagnetic disturbance (GMD) system impact. In *2017 IEEE Power & Energy Society General Meeting*, 1–5. IEEE. doi: 10.1109/PESGM.2017.8273848.

61. Forbes, K. F., St. Cyr, O. C. (2010). An anatomy of space weather's electricity market impact: Case of the PJM power grid and the performance of its 500 kV transformers. *Space Weather, 8(9)*. doi: 10.1029/2009SW000498.

62. Gérin-Lajoie, L., Mahseredjan, J., Guillon, S., Saad, O. (2013, July). Impact of transformer saturation from GIC on power system voltage regulation. In *Proceedings of International Conference on Power Systems Transients (IPST2013)*, 18–20.

63. Overbye, T. J., Shetye, K. S., Hutchins, T. R., Qiu, Q., Weber, J. D. (2013). Power grid sensitivity analysis of geomagnetically induced currents. *IEEE Transactions on Power Systems, 28(4)*, 4821–4828.

64. Zhang, Y., Shetye, K.S., Raymund, H., Overbye, T. (2018). Impact of Geomagnetic Disturbances on Power System Transient Stability. In *2018 North American Power Symposium (NAPS)*, 1–6. doi: 10.1109/NAPS.2018.8600579.

65. Overbye, T. J., Hutchins, T. R., Shetye, K., Weber, J., Dahman, S. (2012). Integration of geomagnetic disturbance modeling into the power flow: A methodology for large-scale system studies. In *2012 North American Power Symposium (NAPS)*, 1–7. IEEE.

66. Pulkkinen, A., Bernabeu, E., Eichner, J., Beggan, C., Thomson, A. W. P. (2012). Generation of 100-year geomagnetically induced current scenarios. *Space Weather, 10(4).* doi: 10.1029/2011SW000750.

67. Overbye, T., Shetye, K., Hutchins, T., Zhu, H. (2013). Resiliency for High Impact, Low Frequency Events (6.1). *PSERC Future Grid Initiative Proceedings*, 21.

68. Overbye, T. J., Shetye, K. S., Hughes, Y. Z., Weber, J. D. (2013). Preliminary consideration of voltage stability impacts of geomagnetically induced currents. In *2013 IEEE Power & Energy Society General Meeting*, 1–5. IEEE. doi: 10.1109/PESMG.2013.6673068

69. Klauber, C., Juvekar, G. P., Davis, K., Overbye, T., Shetye, K. (2018). The potential for a gic-inclusive state estimator. In *2018 North American Power Symposium (NAPS)*, 1–6. IEEE.

70. Shetye, K. S., Overbye, T. J., Qiu, Q., Fleeman, J. (2013). Geomagnetic disturbance modeling results for the AEP system: A case study. In *2013 IEEE Power & Energy Society General Meeting*, 1–5. IEEE.

71. Gérin-Lajoie, L., Mahseredjan, J., Guillon, S., Saad, O. (2014). Simulation of voltage collapse caused by GMDs–Problems and Solutions. In *Proc. CIGRE Session*, 1–10.

72. Haque, A., Vaile, J., Rutkunas, T., Kodsi, S., Bhuiya, A., Baker, R. (2017). Geomagnetic disturbance storm system impact—A transmission facility owner case study. In *2017 IEEE Power & Energy Society General Meeting*, 1–5. IEEE. doi: 10.1109/PESGM.2017.8274485.

73. Hebner, R. (2017). Nanogrids, microgrids, and big data: The future of the power grid. *IEEE Spectrum Magazine*, 23.

74. Babaeiyazdi, I., Rezaei-Zare, M., Rezaei-Zare, A. (2019). Wind Farm Operating Conditions under Geomagnetic Disturbance. *IEEE Transactions on Power Delivery, 35(3)*, 1357–1364. doi: 10.1109/TPWRD.2019.2940913.

# 5 Mitigation of Negative Geomagnetic Disturbance Impacts on Power Systems

## CHAPTER CONTENTS

5.1   Definition and principles of power system resiliency .................................. 121
5.2   Forecasting of geomagnetic activity ........................................................... 127
5.3   Technical actions ........................................................................................ 136
5.4   Operational procedures ............................................................................... 146
5.5   Legislative procedures ................................................................................ 155
5.6   Conclusion .................................................................................................. 161

Prudence requires that actions should be taken to mitigate the risk of GMDs. On one hand, one needs to account for the very high impact of extreme events for the construction of "zero risk power grid". On the other hand, the low probability of such events makes it hard to develop a cost-risk analysis. Moreover, the uncertainty of the network evolution and the change in the events frequency make the picture even more complex. Development of mitigation techniques against GMD requires the knowledge of the full space weather chain starting from understanding the physics of the Sun to the effects on the technological systems. GMD impact on power systems is a complex and technical issue, but many of the potential consequences are common to other risks. Plans to respond to severe GMD can be dealt under existing plans for other events, but in some parts existing capabilities are insufficient. There is no "end-to-end" product which covers the whole spectrum of activities needed for fully operational GMD services. Nevertheless, the set of space-based and ground-based mitigation actions exists. In this chapter, the key challenges and the steps that are crucial for creating future GMD resilient power systems are described that will ensure energy security and sustainability.

## 5.1   DEFINITION AND PRINCIPLES OF POWER SYSTEM RESILIENCY

The power system reliability as a concept of safety refers to the ability to ensure stable operation in case of a common failure appearance, limit number of incidents and consequences to the grid by itself and to the customers. Common failures are caused

by the events with high probability, relatively short duration and relatively small affected region [1]. Traditionally, the probability of whether or not an undesired event
will occur and how severe its consequences would be were given in order to develop
mitigation strategies. Since recently it is believed that there is no inherent probabilities describing the system, and whether events occur or not and how severe their
consequences might be is uncertain, and being dependent on the state of knowledge
[2]. Concept of resilience extends the traditional system limits.

The majority of risk assessments take the approach specified in ISO 31000 [3].
This standard promotes the idea of risk identification as the first step of risk management. Therefore, the common approach includes the process of a series of individual risks identification, assessing their likelihoods and consequences followed
by the comparison of risks based on the severity of impacts. This information is
later used for the National risk matrix construction which is an important mechanism in information sharing within wider community and awareness raising. Only
the highest rated risks are addressed normally. The disadvantage of this approach
is that it fails to assess non-linear/complex risks [4]. Cause-effect relationships for
complex risks can be understood only in hindsight after a disaster occurs. It differs
from linear/complicated risks for which cause-effect relationship can be understood
in advance and stochastic approaches are appropriate for their assessment. It is of
particular importance for the GMD risk as an emerging risk.

Based on the risk type, two resilience building approaches can be distinguished:
specified and general. Specified resilience refers to the known risks, which effects
have been already observed in the past. Risk assessments are based on linear cause-
effect relationships and the strategy follows a "sense and respond" principle [5].
Risks are addressed as "systems of subsystems", since it is assumed that the sum
of composed action plans makes the whole community more resilient [6]. General
resilience refers to the ability of society to withstand unknown shocks. Approaches
to build general resilience have to be simultaneously implemented top-down and
bottom-up.

Resiliency (or resilience) comes from Latin word "resilio" which means "to jump
back" [7]. In 1973, C.S. Holding defined resiliency as a measure to "the persistence
of systems and of their ability to absorb change and disturbance and still maintain
the same relationships between populations or state variables" [8]. After Holding,
numerous interpretations of resilience have been developed. According to the United
Nations Office for Disaster Risk Reduction, resiliency is the ability of a system,
community or society exposed to hazards to resist, absorb, accommodate, adapt to,
transform and recover from the effects of a hazard in a timely and efficient manner,
including through the preservation and restoration of its essential basic structures
and functions through risk management [9]. [10] emphasizes that resilient system is
able to absorb lessons for adapting its operation and structure to prevent or mitigate
the impact of similar events in the future. Resilient system is specified as a system of
**four "R"** [11]:

- **Robust**: strength, or the ability of elements, systems, and other measures
  of analysis to withstand a given level of stress or demand, without suffering

degradation or loss of function. It also includes robust human resources. Satisfying this parameter targets correction of design issues as poorly detailed, improperly restrained or vulnerable.

■ **Redundant**: capacity of satisfying functional requirements in the event of disruption, degradation or loss of functionality. It refers to other measures or systems which are substitutional to existing ones (Eq. 5.1).

$$Redundancy = f\left(\text{reserve capacity}, \frac{1}{\text{time to access}}\right) \quad (5.1)$$

■ **Rapid**: the capacity to meet priorities and achieve goals in a timely manner in order to contain losses, recover functionality and avoid future disruption. In addition, it shows how fast the society can learn from the event.

$$Rapidity = \frac{dQ(t)}{dt} \quad (5.2)$$

$Q(t)$ is a functional level.

■ **Resourceful**: the capacity to identify problems, establish priorities, and mobilize alternative external resources when conditions exist that threaten to disrupt some elements, system, or other measures. It is sometimes thought as how to improve redundancy by providing measures to maintain additional resources and increase rapidity ex post by making investment ex ante.

According to [12], the qualitative analysis of a system's resilience is tighten to three following capacities:

■ **Absorptive capacity** – the endogenous system quality to absorb perturbations in the system with minimal consequences to its proper functioning.
■ **Restorative capacity** – the ability of a system to be easily maintained and dynamically repaired.
■ **Adaptive capacity** – the endogenous system quality to reorganize when facing a shock, so that it reaches its past performance to the fullest possible extents.

Mitigation refers to the measures taken to reduce vulnerability and to increase resiliency of the system. The summary of resilience practices as a function of party involved in support of resilient power grid characteristics is given in Table 5.1 [13].

Power systems during their operations are exposed to a variety of disturbances that result in different ways on the power system behavior. Resilient power grid is operated with an aim to minimize the potential consequences resulting from a disruptive event and to efficiently recover from a potential system performance loss [14]. Depending on the power system capabilities to adapt, self-organize and recover, the system could either collapse to zero or achieve an even better performance after the event. Moreover, a resilient grid should be also able to anticipate, adapt and recover after the event. Twelve resilience-related indicators for power grids are identified in [15].

**Table 5.1**
**Summary of resilience practices to geomagnetic disturbances**

| Robustness | Resourcefulness | Rapidity | Redundancy |
|---|---|---|---|
| **People & Processes** | | | |
| - Extensive continuity of operation plans | - Highly trained and qualified transmission system operators<br>- Cascading failures prevention | - Mutual aid agreements<br>- Priority recovery of critical customers | - Revised emergency response plan after natural and techno-logical catastrophic events<br>- Revised industry standards after no-table blackouts |
| **Infrastructure & Assets** | | | |
| - Interconnected grids provide absorptive capacity<br>- Double redundant transmission links to support N-2 failures | - State-estimators support in real-time monitoring<br>- Automated system transfer for N-1 failure | - Shared inventory of spare high-voltage power trans-formers | - Infrastructure redesign |

Concept of resilient power system targets the creation of the grid resistant to any catastrophic contingency with the various magnitudes. The goal is to smoothen the negative effects of the "threat-specific" risk modeling which leads to duplicated efforts and at times, shortsighted mitigation strategies that result in suboptimal investment decisions [16]. [1] proposes to divide contingencies in two categories: typical power grid outages and super storms. The North American Electric Reliability Corporation (NERC) report [17] also classified GMD as a real high-impact low-frequency risk. In reality, GMD impact on power grid can be better described as a perfect storm. It is shown further that starting from the certain intensity level, CMEs are better predictable, since they require more time to appear. Their characteristics are given in Table 5.2.

Power systems were designed in a way to provide reliable and high quality operation with a reasonable risk level. In other words, algorithms for power system operation are processed in regard with more typical threats. However, events with lower probability may impact power grid resiliency in much stronger way. Even when risk registers and national strategies make the appropriate considerations, the tendency is to separate the categories of risk and to focus on the triggers that are perceived more likely to happen [18].

**Table 5.2**

**Characteristics of typical power grid outages and extreme events**

| Typical outage | Super storm | Perfect storm |
|---|---|---|
| - Disasters severity is proportional to the force | - Multiple forces join together to create a disaster greater than the sum of its parts | - Multiple forces join together to create a disaster greater than the sum of its parts |
| - Forces that create a disaster can be assessed in a systematic way before an event | - Disruptive forces are not assessed before an event | - Forces that create a disaster can be assessed in a systematic way before an event |
| - Almost predictable/ controllable | - Less predictable/ controllable | - More predictable/ controllable |
| - Limited number of faults | - Multiple faults | - Multiple faults |
| - Quick restoration | - Need for more time and resources for restoration | - Need for more time and resources for restoration |

The conceptual resiliency curve shows the resilience level as a function of time with respect to disturbance event [19]. Three levels of resiliency are concerned. The level $R_0$ states for the initial level. System can be considered as a resilient one if the initial resiliency level $R_0$ is high enough to withstand the extreme events. Values $R_{pe}$ and $R_{pr}$ respectively indicate the post-event resiliency level and the post-restoration resiliency level. The level $R_{pr}$ can be the same as $R_0$ ($R_{pr} = R_0$), higher ($R_{pr} > R_0$) or lower ($R_{pr} < R_0$). The post-restoration state is ended at the moment $t_{pir}$. The features of the system such as resourcefulness, redundancy and adaptive self-organization not only define the post-restoration resiliency level $R_{pr}$ but also the time frame of restorative state $t_r < t < t_{pr}$. The mitigation actions should aim at reducing the drag in resiliency levels after the event ($\Delta P_0 - P <_{pe}$) and reducing the recovery time ($\Delta t_{pt} - t_r$).

Quantifying resiliency is required for evaluating the implemented and planned mitigation strategies. It is not a straightforward process, since it is a multidimensional and dynamic process. Resiliency metrics are time-dependent and characterize the grid performance on the moment of initial shock, the function achieved during the event and effectiveness of post recovery.

It is crucial to distinguish between the operational and infrastructure resiliency that are quantified using a different set of metrics. Operational resiliency refers to the characteristics which would help a power grid to maintain operational strength and robustness in the face of an extreme event. Infrastructure resiliency refers to the physical strength of a power grid for minimizing the portion of the system that is damaged, collapsed or in general becomes nonfunctional [1]. For instance, the framework for evaluating grid resiliency by measuring the recovery time and system states is presented by [20]. [21] proposes to calculate the resilience as an index between the real performance and the target performance of the system. In a similar way, [22] compares an improvement of gird functionality from the damaged state with the reduction of functionality from the normal state. The approach of calculating two indices such as grid physical hardiness and operation capability is given by [23].

The growth of population and economic development shifts the energy consumption toward a higher share of electric power. New technologies and associated process development result in higher cost of electricity undersupply. Moreover, the society expects an increased reliability of power supply and a reduced restoration time. Overall, risk management process consists of continuously repeated four phases named as preparedness, response, recovery and prevention. The process is constantly repeated in order to improve the system's performance by using the findings of the previous cycle. Risk preparedness is built closely on prevention. Further, the ability to respond is determined by preparedness level. The recovery phase is often seen as a window of opportunities, since the stakeholders wish to avoid the same catastrophic situation and are more eager to invest in mitigation actions implementation. It is driven by the fact that the event review and impact assessment are conducted at this moment. Normally, it is difficult to justify the high cost prior to the event, since the system can remain reliable under a wide range of ambient conditions by implementing corresponding "defense and restoration plans". Similarly to risk management process, resilience assessment framework is also of four phases. The corresponding phases are: threat characterization, vulnerability of system's components, system reaction and operation and system's restoration [24]. The trap that limits both approaches is that lessons learned after disasters are focused on actions specified in existing regulations rather than the disaster's development and changed conditions [25].

One of the reasons to study system's resiliency is the difficulties in forecasting severe GMDs (see Section 5.2). The extreme contingencies cannot be controlled due to their nature, but their impacts can be managed. The GMD impact on power grids is a complex and technical issue, but many of the potential consequences are common to other risks. Plans to boost resiliency to severe GMD can be dealt under existing strategies for other events, but in some parts the existing capability is insufficient. Resilient grid rather than the reliable grid copes with the extreme events in an active way. There is no end-to-end product which covers the whole spectrum of activities needed for fully operational GMD services. Nevertheless, the set of space-borne and ground-based mitigation actions exist. In general, the following principles for effective mitigation actions planning, as shown in Fig. 5.1, should be applied.

In Fig. 5.1, **Preparedness** indicates the proper preparedness of all relevant stakeholders, including clarity of the roles and responsibilities. **Continuity** shows that all the actions should be grounded within the existing functions and be familiar. **Forecast** stays for adequate forecast of the risks and developing models of direct and indirect impacts. **Integration** supports the parameter **continuity** by saying that

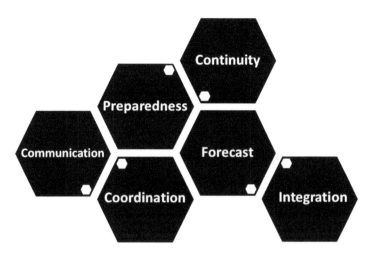

Figure 5.1: The combination of principles for boosting power system resiliency

appropriate guidance should be developed and effective training exercises should be performed. Only the engagement based on the mutual trust brings fruitful **Coordination** and facilities the information sharing. The two way **Communication** is crucial.

The ability to respond to severe GMD strongly depends on the level of preparedness. Only mitigation procedures developed and taught before the event can be quickly used in a crisis. The effectiveness is validated by exercises and real-life events [26]. GMD research and science are a relatively young field. Therefore, the links between science and resilience planning are not mature. It shows that power system resiliency can be characterized as a socio-technical resilience. It is proposed to consider power grid as a network consisting both of physical components and individuals that operate. The power system then can be seen as a network embedded in the national social setting.

The whole further described range of mitigation actions can be divided into four big groups (Fig. 5.2): forecasting of geomagnetic activity (Section 5.2), technical actions (Section 5.3), operational procedures (Section 5.4) and legislative procedures (Section 5.5).

After the 1989 event, the guidelines designed to protect the security of the power system as a whole were proposed [27]. They include: (1) reduce the output of generators to ca. 80% of full load, (2) enable transformer tripping for gas accumulation or sudden pressure relay tripping, (3) remove problematic equipment, (4) adjust loading on HVDC circuits to be within 40–90% range of the nominal rating, (5) reduce loading of inter-area or on critical transmission lines to 90% or less of their nominal rating. The feasibility of their implementation nowadays and overview of the used methods and to which degree they are capable to mitigate the GIMD effects are given further. It was also noted that proposed guidelines may conflict with secure operation guidelines.

Figure 5.2: Complex of measures for preventing negative geomagnetic disturbance impacts on power systems

## 5.2   FORECASTING OF GEOMAGNETIC ACTIVITY

GMD forecasting system is a high-level coordinated system of satellite-based and ground-based facilities to ensure the availability, quality and interoperability of the measurements that are essential for supporting GMD warning. GMD forecasting shares much with forecasting weather on Earth. However, GMD prediction has far to go before it develops to the level of terrestrial weather forecasting in accuracy and lead-time. Wrong information on GMD appearance can lead to human failure, technical solutions misoperation or significant economic losses.

The foremost global authority for forecasting and monitoring GMD is the Space Weather Prediction Centre, which is the part of the US National Oceanic and Atmospheric Administration (NOAA SWPC) located in Colorado, US. The SWPC was initiated in 2005 and exceeded 52,000 customers in 2018. Until recent time, it was the only 24/7 civilian GMD prediction center. UK Met Office established their full 24/7 prediction service in 2014. The Met Office Space Weather Operations Centre (MOSWOC) coordinates with NOAA SWPC. It is an essential step for GMD forecasting quality improvement, since space weather phenomena are best forecasted through coordinated efforts of multiple nations.

The types of instrumentation used for monitoring and exploring space weather environment have been quite stable since the beginning of the space era, though sensor technologies, electronics, and processing technologies were significantly improved in the course of past decades. Radiation belts were discovered by James Van Allen using the right instrumentation in the right place – and most importantly – by correctly interpreting unexpected measurements. The discovery was made by instruments intended to measure cosmic rays aboard Explorer I that was launched at 03:48 UTC on February,1, 1958. The cosmic ray instrument was an Anton type 314 Geiger-Mueller (GM) counter with an omnidirectional geometric factor $G_0 = 17.4$ cm$^2$, a total (stainless steel) shielding of 1.5 $g/$ cm$^2$, corresponding to thresholds of 30 MeV for electrons. The unexpectedly high and zero counts were correctly interpreted by James Van Allen and Carl McIlwain as the result of much higher fluxes than the expected cosmic rays. These high fluxes saturated the GM counter electronics. Post-flight tests on a spare instrument confirmed these results.

GM counters were the first instruments aboard spacecrafts at the birth of the space age. They operate by detecting energetic particles incidents in a tube filled with low pressure gas. The particles ionize the gas, and the ions are accelerated in the electric field between the wall and the axial wire. In the GM voltage regime, the accelerated ions create an avalanche that drops the voltage (short-circuit). This short-circuit appears as a negative voltage pulse at the output. The output pulse is independent of the particle's species and its energy, as long as the particle penetrates into the tube. The threshold energy of the particles detected can be defined by using shielding of different thickness around the sensitive volume of the tube.

The Explorer IV satellite launched on July 26, 1959 was instrumented by James Van Allen and his team at Iowa university for observing the Earth's energetic particle environment. The instrumentation comprised two scintillation detectors and two GM counters (type Anton 302). The Explorer IV satellite project was undertaken in conjunction with operation Argus, a military project to test the feasibility of creating an artificial radiation belt of high energy electrons by a high altitude nuclear explosion [28]. The objectives of the mission were to confirm the existence and configuration of the natural radiation belts and to assess the formation of the artificial radiation belt. Both objectives were achieved.

The ATS-1 satellite launched on December 6, 1966 was the first satellite in geosynchronous orbit to carry radiation detectors. It provided the first opportunity to observe the large- and small-scale temporal variations in the trapped electrons and protons in the region beyond the maximum in the outer radiation belt and the boundary of the magnetosphere. The first optically observed CME was detected on December 14, 1971 by R. Tousey using the seventeenth Orbital Solar Observatory [29]. CMEs are best seen with white-light coronagraphs that create and artificial solar eclipse and record sunlight that has scattered from coronal electrons. Nowadays, space weather forecast is based on chronographic observations received from the following European Space Agency (ESA) and the National Aeronautics and Space Administration (NASA) missions:

1. The STEREO (Solar TErestrial RElations Observatory) mission of NASA consisted of two nearly identical space-based observations – one ahead of Earth in its orbit (STEREO-A), the other trailing behind (STEREO-B) to provide stereoscopic observations of the Sun and in-situ measurements for

the study of the solar wind and its dynamic features such as Stream Inter-action Regions (SIRs) and CME. It was launched on October 25, 2006.

2. Aged ESA/NASA SOHO (Solar & Heliospheric Observatory) mission was launched in December 1995 to study the Sun, from its deep core to the outer corona, and the solar wind. SOHO moves around the Sun in step with the Earth, by slowly orbiting around the Lagrangian point (L1) point. While SOHO continued operation into the 2020s depends only on the longevity of its solar arrays, there is as yet no defined mission to succeed it in providing continuous, earth-Sun-line coronagraph observations [30].

The advanced space weather forecast is still more a nowcast. The success of long-lead time forecasting depends on predicting accurately intrinsic properties of a CME when it erupts from the Sun, and on how intrinsic properties change during the prop-agation from Sun to Earth [31]. The propagation of CME is normally covered by modeling which is sensitive to CME properties evolution along their way [32]. In several cases, an observation of a fast and bright Earth-directed CME may result in quiescence: CME points to north [33], CME is channeled away from the Sun-Earth line [34], some CME leave the Sun without noticeable features [35], CME originat-ing close to the limb of the Sun may deflect toward Earth [36], etc. These false alarms are a major concern of space weather forecast users. Space weather forecasting by itself is highly reliant on ongoing scientific research, including model development, as well as mainly relying on scientific instrumentation that is not optimized for real-time usage [37]. Recently, NASA launched the Parker Solar Probe in August 2018 for improving our understanding of the sun's corona and the origin and evolution of solar wind.

In total, four types of messages exist for describing GMD status. The example of their visual representation is given in Fig. 5.3 seen at SWPC facilities. The example of the customized GMD visualization tool for real-time information of GIC flows at key locations in the system is given in [39]. The tool, developed by Dominion Vir-ginia Power (US), maps NOAA's forecasts to reflect its risk to an impending GMD event. Based on the storm forecasts and local GIC sensor measurements, system operators achieve different levels of control, including recording storm events, mon-itoring system voltage profiles, monitoring susceptible GMD-influenced locations in Dominion Virginia Power service territory, monitoring generation step-up trans-former and transmission transformers for potential overheating conditions, adjusting equipment outage and maintenance schedules, and preparing possible generation re-dispatches and load-shedding plans in response to severe GMD scenarios [39]. The procedures are compiled with PJM's Manual 13 – Emergency Operations.

1. **Watch**, issued as soon as an Earth-directed CME is detected by SOHO and STEREO missions. On average, transmission system operators (TSOs) receive information 48 hours in advance of a predicted GMD that may cause power system state instabilities.

2. **Warning**, when a CME is detected by ACE/ DSCOVR satellites about 15–60 minutes before impacting Earth. This is the moment when magnetic

field orientation can be measured. Since Earth's magnetic field points north, southwards oriented CME may have considerable impact on power grids. Northwards oriented magnetic field will mainly impact satellites and aircrafts (see Section 2.1). ACE (Advanced Composition Explorer) is a NASA mission launched in 1997 to study solar wind physics. DSCOVR (Deep Space Climate Observatory) is a NASA mission launched in 2015 to succeed ACE in supporting solar wind alerts and warnings from the L1 orbit.

3. **Alert**, issued when the GMD is detected at magnetic observatories (see Chapter 3). It is almost a near real-time indication that a GMD is occurring.

4. **Summary**, issued after the event ends, and contains additional information that was not available at the time of issue.

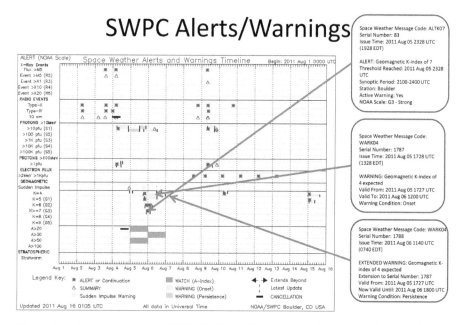

Figure 5.3: Representation of SWPC Alerts/Warning for August 2011 coronal mass ejections (Courtesy of PJM Interconnection)

The time of CME detection by ACE depends on the CME strength. It can shrink to 10 minutes. A 10 minute detection time does not equal to 10 minute action time. It takes several minutes more to proceed and validate the information, deliver it to infrastructure operators. The actual time power grid operators have maybe as little as 5 minutes, which is too small for any meaningful mitigation actions. Another difficulty is that SOHO and ACE solar wind instruments are affected by severe space weather conditions. [40] defines three time frames for space weather forecasting: "no lead time", "short-time lead", "long-lead time". SOHO fails to provide near-real-time information due to its susceptibility to strong solar proton events. However, ACE is unaffected by solar proton events. Its performance degraded during the Halloween event in 2003. It is noted that a remarkably fast CME impacted sensors on ACE. The plasma instrument was disabled and the determination of the shock was solely based on the magnetic field instrument [41].

SWPC forecast is mainly used for situational awareness, both for decision making and planning activities and for understanding perturbations or anomalies in system behavior [42]. In other words, forecast information is used for reasons of human safety, ensuring infrastructure safety, morbidity reduction. Successful forecast of the magnitude and duration of GMD is a silver bullet for efficient GMD mitigation. It is stated that the net benefits of a satellite warning system are extremely positive even if the damage is as low as USD 2 billion [43]. ESA SSA program performed cost-benefit analysis of improved space weather forecasting services ("Do ESA SSA Programme") [44]. The impact of three-day long G5+ scenario on three average European cities (geographically nonspecific) with the current and improved ESA SSA capabilities was modeled. It showed the impact of Euro 2,656 million. The gross domestic loss for UK in case of 1-in-100 scenario is Euro 15.9 billion, reducing to Euro 2.9 billion with current forecasting and further reduced to Euro 0.9 billion with improved forecasting [45]. Actions for improving forecasting capability include:

- coordinated development and validation of end-to-end models and applications;
- ensured availability of the measurement data;
- development of new capability through international collaboration, i.e. combination of data obtained in Lagrangian points: L1 and L5.

The idea of space observation in L5 point originates from the fact that a spacecraft located in L5 point will pass over the same region of the solar surface as the Earth about 4–5 days before the Earth [46]. In addition, L5 is a very attractive point for in-situ scientific measurements that can clarify some of the unknown questions about CMEs [47]. The examples of such missions include: Earth-Affecting Solar Causes Observatory (EASCO) [48], Carrington [49], Instant [50]. Apart from their primary role, STEREO satellites gave an example how the L5 measurement could be used, since STEREO-B crossed the L5 in late 2009. Cross-correlation of data from L1 and L5 was performed in [51]. According to ESA SWE chosen plan mission will be in operation in 2024.

Semi-empirical models are mostly in use for space weather forecasting which are quick to use and computationally efficient. Examples of semi-empirical models are [52], [53] for CME magnetic structure and orientation, [54] CME's altered trajectory and Helicity-CME [55]. Different methods are used to predict the geomagnetic activity indices distribution. Linear regression methods are used in [56], [57], [58], whilst regression techniques for data sets with asynchronous observations are presented in [59]. Neural networks for space weather forecasting are tested in [60], [61], [62], [63]. Artificial intelligence algorithms and machine learning are used in [64], [65], [66], [67], [68]. An attempt to adapt Kalman filter approach was done by [69] and implementation of econometric methods was shown by [70]. These efforts were helpful for the advancement of space weather forecast errors. It is proven to be a difficult task for many areas that space weather forecast includes error bars considering many inaccuracies involved in obtaining observations [71].

In the 25 years of solar wind observation, only 1–5% of time periods refer to intense CMEs (less than 150 intense CMEs were detected). Solar flare forecast may be currently provided to end users as-is, for example, a statement will be issued that there is a 10% chance of flares occurring in the next 24 hours [72]. It is rare to encounter error bars included with the forecasts ($10 \pm 1\%$). One of the main strengths of ensemble forecasting system is their ability to represent the uncertainty that is inherent to any forecast [73]. Examples of existing space weather service are given in Table 5.3 (based on [177]). First general purpose services are listed that provide data and predictions on space weather conditions. The data is delivered in scientific terms, therefore the customer outside scientific community needs expert support. Then the services adapted for power grid specialists are presented.

Many industrial sectors affected by space weather would like to receive the forecast on $B_z$ amplitude within the 24 hours before the event. Moderate and severe events are also of interest to the customers, not only extreme. After the initial eruption, forecasters can map the event to Earth-arrival using a procedure shown in Fig. 5.4. Achieving this is associated with the set of difficulties. The main one is the orientation of the magnetic vectors within CMEs. A south directed solar wind magnetic field is more likely to heat the Earth's magnetosphere, causing the field lines to break and reconnect while releasing energy and causing geomagnetic storms [52]. Magnetic reconnection allows energy to enter the magnetosphere on the dayside, consequently, energy is stored and explosively released on the nightside. A model capable of predicting magnetic vectors with a lead time of more than 24 h is described in [75]. The other difficulty is that aforementioned SOHO and STEREO satellites are scientific missions, therefore scientific measurements and investigations have the higher priority. SWPC specialists control on GOES and DISCOVR satellites.

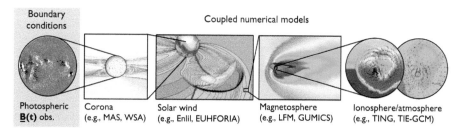

Figure 5.4: A pathway of how numerical simulations can be coupled together from the Solar surface to the Earth [76]

Space weather forecasts are considered as inadequate by 75% of responders, while only about 20% responders rated the forecasts as fully adequate. COSPAR together with ILWS prepared a roadmap for advancing space weather forecast. Improving space weather forecasts requires the enhanced study of the full chain that connects

**Table 5.3**

**Examples of recognized space weather prediction services**

| General purpose services | |
|---|---|
| Space Weather Prediction Centre (SWPC), US | www.swpc.noaa.gov |
| International Space Environment Service (ISES) | www.spaceweather.org |
| Space Weather European Network (SWENET), ESA | swe.ssa.int/TECEES/swenet.html |
| Bureau of Meteorology (BoM) Space Weather Service (SWS), Australia | www.sws.bom.gov.au |
| Geomagnetic Indices Forecasting and Ionospheric Nowcasting Tools, Italia | gifint.ifsi.rm.cnr.it |
| International Service for Geomagnetic Indices (ISGI), France | isgi.latmos.ipsl.fr |
| Norwegian Centre for Space Weather (NOSWE), Norway | spaceweather.no |
| IZMIRAN Space Weather Prediction Centre, Russia | spaceweather.izmiran.ru/eng/index.html |
| **Power grid** | |
| GIC Now! | aurora.fmi.fi/gic_service/english/index.html |
| GIC Forecast | www.lund.irf.se/gicpilot/gicforecastprototype/ |
| Real-Time GIC Simulator | www.spaceweather.gc.ca/tech/se-gic-en.php |
| Solar Wind Monitoring and Induction Modeling for GIC | www.geomag.bgs.ac.uk/gicpublic |

the Sun to society. The roadmap team identified the highest priority areas within the Sun-Earth space-weather system, which advanced scientific understanding is urgently needed to address current space weather service users requirements (Fig. 5.5) [167]. Figure 5.5 focuses on CME post-eruption. The roadmap recommends coordination in addressing key science challenges:

- data need (both ground and space);
- smooth transition of scientific developments into reliable services.

Space weather services have not reached a level of maturity comparable with

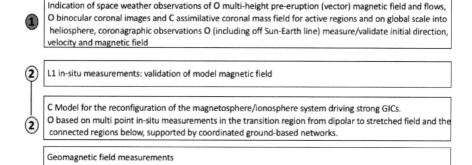

① Indication of space weather observations of O multi-height pre-eruption (vector) magnetic field and flows, O binocular coronal images and C assimilative coronal mass field for active regions and on global scale into heliosphere, coronagraphic observations O (including off Sun-Earth line) measure/validate initial direction, velocity and magnetic field

② L1 in-situ measurements: validation of model magnetic field

② C Model for the reconfiguration of the magnetosphere/ionosphere system driving strong GICs. O based on multi point in-situ measurements in the transition region from dipolar to stretched field and the connected regions below, supported by coordinated ground-based networks.

Geomagnetic field measurements

Figure 5.5: Top priority needs to advance understanding of space weather to better meet user needs. Letters identify opportunities for immediate advance (O) and the largest challenges (C)

other meteorological services. Hence, potential users may not be aware of the capabilities and benefits of space weather services and how to use them. Alerts, warnings and forecast should be communicated effectively during extreme events. Stakeholders are often reluctant to devote time and effort to understand how to interpret space weather data and may be confused when large amounts of data are provided [78]. Stakeholders prefer to receive the data that gives clear correlation with GMD conditions and the possible impact on the grid. In other words, stakeholders prefer to receive "action-based" forecast rather than the information on devoted variables amplitudes. For example, Food and Agriculture Organization applied Early Warning Early Action system (EWEA) [79]. This experience should be adopted by other industries.

The classification of risk indicators can be developed using the example of the tools already developed by the Community Coordinated Modeling Center at NASA (https://ccmc.gsfc.nasa.gov). Space weather services to TSO must be standardized and harmonized. Furthermore, stakeholders prefer to receive the complete data including the information on the interplanetary magnetic field orientation earlier than it is possible nowadays. Especially since the wrong information on GMD appearance can lead to human failure. Decision making process is normally based on the trade-off between event probability and lead-time. Instead low confidence early warning can be distributed among the stakeholders which is followed by an almost precise forecast at L1 point.

The practice of ensemble techniques is common in terrestrial weather forecasting. This allows stakeholders to adjust the technological systems characteristics and still gives the option to cancel control actions if they are no longer needed. The visualization of the technique is presented in Fig. 5.6 (this figure is adapted from [153]). The heavy lines represent the evolution of the single best analysis of the initial state (deterministic forecast) and dashed lines – the evolution of individual ensemble member (probabilistic forecast). The examples of uncertainties are uncertainties in the state of the heliosphere leading to uncertainties in the state of the Earth's magnetosphere and thermosphere, ionosphere and ultimately in the predictions of impacts on satellite navigation signals, geomagnetically induced currents and satellite drag [80]. In practice, real ensemble prediction systems may not capture the whole spectrum of uncertainties.

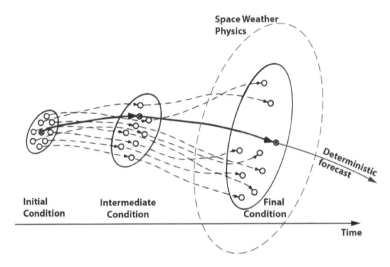

Figure 5.6: Schematic illustration of some concepts in ensemble forecasting, plotted in terms of an idealized two-dimensional phase space

The characteristic example of the practice of space weather forecasting for power grid operation is Hydro-Québec power grid. Hydro-Québec first signed an agreement with the Geological Survey of Canada for providing space weather forecast on 24/7 basis. Technical mitigation actions, which are the topic of the next sub-chapter, were implemented afterwards.

## 5.3 TECHNICAL ACTIONS

Past events showed that GMDs do pose a serious risk on power grid operation. However other terrestrial hazards threaten grids more often, which results in better developed practices for their mitigation. GIC prevention measures can be described as infrastructure hardening in nature. This section describes the set of technical actions aiming at GIC effects mitigation. It has to be assured that any action taken for GIC mitigation does not endanger the normal power grid operation. This is normally reached through the use of electric power industry standards and the testing of equipment against such standards. Currently, there is a limited operational deployment and testing of GIC-blocking equipment with a large amount of GIC.

The first group of actions redirects GIC flows over network element either by installation series capacitor in transmission line or DC blocking device in power

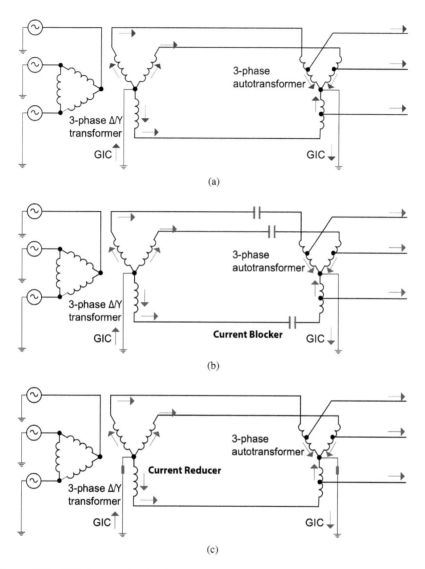

Figure 5.7: GIC flow paths (a) base case (b) in the presence of transmission line series capacitors (c) in the presence of DC blocking devices in transformer neutral (Courtesy of Metatech Corporation)

transformer neutral. The first approach is a proactive strategic vision for improving power grid resiliency not only in the presence of GIC, while the second is reactive (Fig. 5.7). Series capacitors installed in the transmission line block the GICs flows, which are quasi-direct in its nature. It needs to be applied in all three phases of the transmission line, which increases the action cost. Series capacitors are normally installed in the long high-voltage transmission lines for improving angular stability. By reducing the overall reactance between the line ends, the flow of active power is, consequently, increased. The power flow $P$ across the line is defined as in Eq. 5.3.

$$P = \frac{U_1 \times U_2}{X_L - X_C} \times sin\delta, \tag{5.3}$$

where $U_1$ and $U_2$ are end voltages of the transmission line, $X_L$ is line reactance, $X_C$ is reactance of the series capacitor and $\delta$ is an angular separation between the line ends.

Another network level solution is an underground HVDC power grid or a space power grid. HVDC underground power grid project involves installation Elpipes in steel underground pipes in support of existing power network for reinforcing them [81].

It is proven that the weak link in power grid operation in the moment of GIC appearance is the power transformer (see Chapter 4). Since their vulnerability varies with the construction type and isolation characteristics, it is not possible to recommend a universal applicable protection method. Nevertheless, general recommendations are given. Since the way for GIC to penetrate inside the grid are the grounded neutrals, the most obvious solution is to disconnect neutral wires. Hence, opening the neutral results in safety and insulation problems. Replacing solid earthing of power transformers with a capacitor or a resistor in a neutral targets the protection of specific transformer unit and power flow redirection.

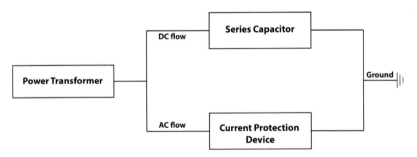

Figure 5.8: Logic scheme for DC-blocking device

The series capacitor installation and neutral point reactors is not a straightforward approach to mitigate GIC risks. GIC blocking devices in neutral consist of capacitive or resistive circuits placed in the ground connection of the transformers [82]. Capacitive device blocks the quasi DC flow, but do not impact the normal 50/60 Hz AC flow. The device entails two primary circuits: a circuit with a blocking device for GIC capturing and a bypass circuit for a AC flow in case of a short-circuit mitigation (Fig. 5.8, adapted from [96]). Once the fault has been isolated from the network and AC current level drops, the device automatically reverts to the normal GIC blocking mode of operation [83]. A circuit with a blocking device can be modeled as an open switch between the substation ground and transformer neutral [84]. The choice of parameters has to be done while considering ferro-resonance issues and potential risk of impedance change elsewhere in the grid. Another concern is the DC-blocking device impact on distance relay protection operation [85]. Moreover, they should not call for more intervention than the transformers to which they are connected. The detailed list of device performance requirements is given in [86]. It is especially important to determine the magnitudes and locations of GIC for allocating the installation places due to the budget constraints. The GIC level highly depends on the geoelectric field direction and can vary from $\sim 0$ A to its maximum value. The solutions for optimal DC-blocking device placement are presented in [87] using branch and cut algorithm and in [88] by considering equipment thermal limits. Each blocking capacitor or resistor costs about half a million including sensors and labors [89], [90]. Since the resistance of a reactor is typically 2–4 times bigger than the other contributions to the resistance which determines GIC, it is presumed that the reactor installation can be an effective measure. The results show that the installation of a reactor at almost a half of all stations makes GIC smaller at these particular stations, but the overall average decrease of GIC in the system is not very large [25], [89]. The reability function on the number of blocking devices can be expressed as in Eq. 5.4 [93].

$$\sum_{i\in(N\cup S)} z_i \leq P, \tag{5.4}$$

where $z_i$ is a binary variable which indicates whether the node $i$ has DC-blocking device or not; $P$ is the maximum number of blocking devices defined by the grid operator/owner based on the economical constraints.

Another approach is the installation of low-Ohmic resistors in the grounding wire. These resistors are relatively cheap in production, since they need to withstand only low voltage and current ratings. The resistor in neutral does not eliminate the GIC, although it can substantially reduce it at the cost of loss in protection sensitivity and size of the equipment [83]. Appropriate use of them does not interfere safety guidelines of power grid operation. Approaches for adaptation of non-linear resistor (either as a metal-oxide varistor or a surge arrester) to mitigate undesired GICs were advanced in [94] and [95].

A complimentary approach is the inclusion of sufficient "grounding" points to allow the induced currents non-destructive paths to ground from the transmission lines. Typical requirements say that the 50 km interval in North-South direction and 75 km interval in East-West domain should be provided. Closer to tropics, interval length increases up to 60 km. While this strategy lessens the chances of GICs being carried to critical equipment, complete protection is not guaranteed [97].

The example of industrial solution aimed at hardening grid from GIC is the "SolidGround" GIC stability system (Fig. 5.9, adapted from [100]) developed by the Swiss company, ABB. The device provides a metallic ground path, which solidly grounds power transformers in the absence of GMD, and triggers removal of the metallic grounding by leaving the neutral grounded though a $1\Omega$ resistor and 5,600 kVAr capacitor bank in the presence of GIC. The control electronics incorporate amperage, harmonic content and time delay through user-configured thresholds. The choice between automatic and manual modes is offered. Under normal operating conditions, presence of this protection is completely transparent to grid operations as the transformer neutral has a complete metallic path to the ground [99]. For 99.8% of time the device is operated through the metallic ground. Protection mode through resistor/capacitor bank corresponds to 0.2% of time. Surge arrester handles the rare events of a simultaneous DC current and a ground fault.

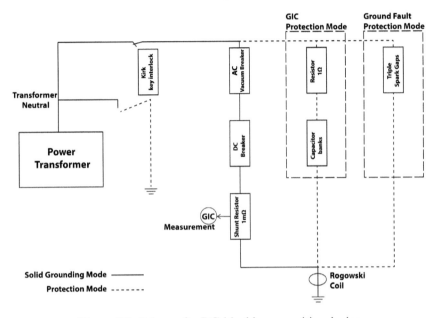

Figure 5.9: Scheme for DC-blocking capacitive device

The following papers made a contribution in showing that switching can mitigate GIC effects [101], [102], [103]. Contrary to DC-blocking device installation, line switching does not involve any additional cost. GICs as another form of a contingency may be also effectively mitigated by a corrective line switching. Line switching modifies current flows over the network elements in the way to relieve line overloads and voltage violations. Topology control in power system operations is a rather old technique used to tackle power grid operation issues [104], [105]. Advanced approach which also implies the whole variety of operational procedures is given in [106]. More information on operational algorithm adaption is given in Section 5.4.

Under anomalous magnetic flux patters, power transformers experience heat dissipation in the parts where heating does not normally occur. It necessitates a power transformer's thermal assessment. The Federal Energy Regulatory Commission (FERC), US, approved NERC's second-stage reliability standard in September 2016. NERC planning standard (NERC Standard TPL-007) asks certain owners and operators to perform network analysis and transformer thermal assessment. Electric power suppliers subject to the standard include transmission owners and generator owners with facilities that include one or more power transformers with a high-side wye-grounded winding with voltage greater than 200 kV and planning coordinators and transmission planners with such facilities in their planning area. Overall, the corrective action plans and assessments have to be in place by January 1, 2022. In particular, Requirement 6 (R6) stays for conducting a thermal impact assessment if $GIC_{max}$ is 75 A or more by January 1, 2021. It should be ensured that all high-side wye-grounded power transformers 200 kV and higher will not overheat during a benchmark GMD event.

In case a manufacturer performs GIC thermal assessment test, the situation seems to be easier, though limited information is available regarding the assumptions used. Hotspot temperatures are strongly dependent on GIC history (moment of appearance, amplitude, duration). The GIC square pulse waveforms vary between 2, 10, or 30 minutes in duration. In plus, they depend on the bulk oil temperature due to loading, ambient temperature and cooling mode. These thermal capability curves are unique and should be developed for every transformer design and vintage. An example of thermal capability curves as a function of GIC amplitude and duration is given in [107]. IEEE C57. 91-2011 (IEEE Guide for Loading Mineral-Oil-Immersed Transformers and Step-Voltage Regulators) specifies thermal limits for cellulose in the paper-oil insulation, which result in its accelerated aging or the potential generation of bubbles in the bulk oil. This information can be used as a basis for a thermal assessment in the absence of manufacturer specific information.

Transformer hot-spot heating is not instantaneous. The hotspot thermal transfer functions can be either obtained from measurements or given by manufactures. IEEE C57.163-2015 Guide for Establishing Power Transformer Capability while under Geomagnetic Disturbances defines the procedure of power transformer thermal assessment. It asks:

- to identify the transformer magnetizing current for the given GIC levels;
- to identify the peak magnetization current and reactive power consumed by the transformer as a function of GIC level;
- to identify the top clamp, tie plate and winding hotspot temperatures;
- to construct thermal capability curves for base and peak GIC levels.

The example of customized thermal capacity curve creation according to the standard is given in [108].

The real-time temperature and dissolved gas monitoring in power transformers provide insight into their condition. The oil in large transformers is normally checked, as part of routine maintenance. The existence of abnormal chemicals and gasses are checked. They are created as a result of oil contamination, insulation breakdown and internal arcing. The used actions include gas-in-oil analysis, acoustic partial discharge detection, moisture sensor, tap-changer operation supervision and others. In parallel, the gradual replacement of conventional cellulose isolation, which is vulnerable to GIC effects, was started 30 years ago. One of the possible solutions is polymeric isolation Nomex [109]. Whereas the polymeric materials are not widely used because their application is associated with the polymeric degradation in oil isolation. Diffusion of small polymer particles into the oil changes the electric properties of oil as an insulation material.

Along with increased stakeholder awareness, any industry develops solutions for GIC effects mitigation with a special focus on power transformers. Power transformer redesign is one of the options. Evident solutions – like increased core cross section or the air gaps inclusion in the core – have significant disadvantage apart from a higher cost. First of all, it is not possible to compensate extremely limited range of GIC amplitudes. Secondly, power transformer's noise level tends to be increased in the construction with air gaps. In the course of scientific evolution for GMD effects mitigation, several exotic solutions were proposed. One of them is an introduction of the same amount of DC ampere turns with opposite sign into a compensation winding, which prevents power transformers from saturation [110]. The challenge of accurate DC magnetization measurement together with the need for low power feed prevents this mitigation action broad implementation. Instead of the air gaps inclusion in the core, it was also proposed to fill the gap with a liquid that changes the total magnetic permeability of the magnetic circuit. Another type of the protection of high-voltage power transformers is made using a special relay containing no microelectronic components and is based on discrete high-voltage elements resistant to electromagnetic interferences and surge over-voltage [111]. The relay consists of a reed switch, RS, with a coil placed on the grounding cable (bus), and a conventional toroidal current transformer installed on the same cable.

Measurement transformer unstable operation can decease power system controllability. However, it was shown in Section 4.1.3 that measurement transformers are relatively robust to GMD effects. Finish experience shows, for instance, that filling a small gap with insulating oil prevents current transformer saturation. In other words, the current transformers feeding the overcurrent relays are usually linearized [123].

Proactive approach for increasing power resiliency to GMD is the planned replacement of power transformer with less vulnerable types. Three-phase three-limb transformer has the highest robustness to GMD. The severe GMD of 2003, so-called Halloween blackout, resulted in power blackout that affected 50,000 customers (Section 2.3.1). During this event, high GICs were registered at Oskarshamn nuclear power plant, Sweden, as well as in 2000. E.On and Fortum who jointly own Oskarshamn nuclear power plant ordered ABB to design and produce the three-phase 825 MVA generator step-up transformer for 400/21 kV. The order was delivered in 2006. The unit contains three-limbs instead of the usual five limbs design, which is common for this power and voltage ratio. It is one of the largest and heaviest cores that ABB has ever produced. It consists of 44,000 steel plates of 1 m wide, 0.27 mm thick which are placed in two stacks. The resultant core's weight is 200 tons [113].

Although utilities typically hold an inventory of spare transformers, the quantities may not be compared against potential needs. Since the transformer's replacement is logistically challenging, US Board on Energy and Environmental Systems recommended instigating a stockpile of easily transported high-voltage recovery transformers in 2012. NERC's Severe Impact Resiliency Task Force identified among others the following action: "Consider the spare equipment critical to BPS restoration and ways to improve the availability of these spares" [14]. In 2014, the Electric Power Research Institute reported the outcome of the Recovery Transformer or RecX project, which was successfully designed, manufactured, tested, transported, installed, energized and field tested a set of rapid deployment high-powered 345 kV emergency spare transformers [115]. The test included a 25-hour journey from St. Louis (temporary storage site at ABB) to a CenterPoint Energy substation in Texas and transformer's installation/ energization in less than 6 days (106 hours). In parallel, the Edison Electric Institute is developing a Spare Transformer Equipment Program (STEP) since 2006. STEP represents a coordinated approach to increase the inventory of spare transformers and optimizing the process of transferring transformers to affected utilities. Essential elements of STEP include: review of present stock levels; transportation logistics plans, contingency mitigation algorithms, spare storage sites protection. The minimum target level of spare transformers should be minimum one spare transformer for each type (voltage level, impedance, construction scheme) unless a range of applicability of a particular spare can be extended to cover several configurations. In addition, NERC Space Equipment Database supports an industry on wide identification of spare equipment. The program is currently voluntary.

Apart from observation purposes (see Chapter 3), GIC monitoring systems are used to visualize GICs in power transformers. The Electric Power Research Institute established a monitoring network called SUNBURST. It automatically correlates GIC presence and measured high harmonic distortion. By the moment of March 1991 GMD, the equipment was installed in four substations, i.e. Brighton, Maryland; Pleasant Valley, New York; Chester, Maine; Deans, New Jersey. The main finding was that placing widely dispersed monitors would preclude gaining warning times. It also suggested that the heating effect could be cumulative as it was later proved by observations in South Africa. The American Electric Power approach for GIC modeling, monitoring and mitigation was presented in [117]. National Grid Company, England, developed an operating application in 2000 that combines data received from ACE and GIC registration. Since its installation the forecasting system provided "watch" and "warning" alarms to the control rooms staff. The most useful forecast displays were found to be the 72 hours ahead, 24 hours ahead and the 1 hour ahead displays [118]. The visualization of the phenomena observed by a transmission system operator is given in Fig. 5.10, where arrows stay for the computed electrojets and the size of the circle is correlated with GIC intensity in the node. Later in 2012, Hydro-Québec and Alstom adopted their Smart Grid technology. In-depth observation and full power flow controllability give higher situational awareness of the grid condition. The power flow can be predictively redirected from susceptible nodes before the disturbance occurred. One more example developed in DominionVirginia Power, US, was described in the previous section.

FERC order on "Reliability Standards for Geomagnetic Disturbances" requires a plan from bulk power systems owners and operators track that instability, uncontrolled separation, or cascading failures of the bulk power system, caused by damage to critical or vulnerable bulk power system equipment, or otherwise, will not occur as a result of a GMD [119]. Significant efforts have been made to incorporate GIC into the power grid modeling algorithms. Several tools have been developed. Despite the rapid development, they are far from the maturity compared to others and the scientific, and engineering gaps still exist. Siemens AG produced **PSS E** power flow network data for GIC calculation [120], which models the set of GMD scenario, variety of conductivity maps, five transformer winding configurations (two-winding transformer and autotransformer, three-winding transformer and autotransformer, phase angle regulators). PowerWorld proposed a simulator which can automatically scale the magnitude according to the parameters of NERC benchmark, which gives power engineers and information of GIC impact and mitigation options [121]. Recently, an open-source MATLAB-based GMD package called MATGMD was introduced [122]. Overall advanced capabilities to model transformer heating and harmonics are not yet available, features of commercial software packages are common. Their application is limited to specific R&D projects.

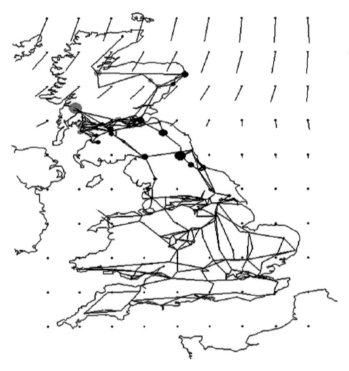

Figure 5.10: Visualization of the April 2000 GMD as observed by National Grid Company operator in the control room [128]

The development of mitigation actions is a challenged for both engineers and economists. Any mitigation action search is supported by economic analysis of their efficiency. As the illustration of technical actions for GMD mitigation implementation, three examples are presented. The first two are real cases: the Hydro-Québec grid hardening after the March 1989 blackout and the transformer current blockers used in New Zealand. The third one is the theoretical study for reducing the GMD risk in the National grid company, UK.

Hydro-Québec is characterized by its unusually high-voltages such as 735 kVAC. More than 1000 km of transmission lines are stretched in the north-south direction. Generation sources are mainly represented by hydro power plants, which are located in the North, far from the load centers in the southern part of the grid. Given its unique geography, Hydro-Québec pioneered in usage of ultra-high voltage transmission lines [123] together with protection techniques [124].

After the 1989 blackout, Hydro-Québec installed a four-point voltage asymmetry measuring system which was later extended to eight points [125]. The system triggers an alarm when defined asymmetry limits are reached. In addition, modern customized PMUs that include algorithmic features specific to Hydro-Québec were installed in 2004. They can register harmonic distortion up to the 10th component. These variables are required by the Hydro-Québec preventive control schemes against GMD induced contingencies [126]. Static VAR compensators normally cover reactive power deficit tripped in 1989 (see Table 2.2). Without the SVC support the voltage declined – "automatic load shedding was not enough to compensate", and the system collapsed within 90 seconds [127]. Consequently, the thresholds were changed.

Nevertheless, the power grid adaptation was primarily driven by stability ensuring reasons. A set of series compensators were installed on the transmission lines heading from James Bay and Churchill Falls [129], and corresponding decision was taken before the event. The power grid map is presented in Fig. 5.11. The retrospective shows that the series compensation effectively blocks the GIC distribution. Accordingly, harmonic distortion is also reduced. For instance, asymmetry grew only until 7.5% at Chateauguay and 5.5% at Micoua on April 7, 2000 [161]. After these steps had been implemented across the grid, according to Canadian organization "there have been very intense GMDs after 1989, but they have not caused any problems" [97], [128].

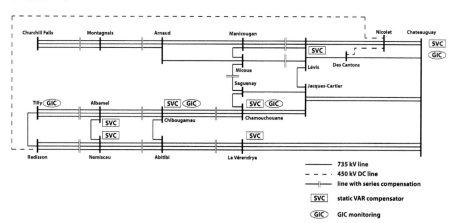

Figure 5.11: Equivalent Hydro-Québec power grid

New Zealand power grid is another example of multi-purpose use of mitigation equipment. Neutral damping devices, which consisted of current measuring devices and neutral earthing devices, were installed and in use since 1992, and neutral blocking devices since 2012–2013, to reduce potential damage due to stray DC currents caused either by HVDC line operation or GMD [131]. Hardware-based protection (both capacitive and resistive devices) is coordinated with the emergency operating procedures and the system restoration plan. Currently, 52 nodes are monitored. The collected data does not only represent history of transformers performance, but also can be used to assess the level of transformer saturation.

National grid is one of the world's largest investor-owned energy companies, and the operator of one of the world's largest privatized high-voltage electric transmission systems at 400 and 275 kV. The UK electricity supply system has already experiences significant GMD effects which are described in details (see Section 2.3.1). Detailed examination of various storm incidents was carried out including (a) SSC; (b) moderate, but frequent electrojet; (c) probable electrojet; (d) severe electrojet, for the both northern and southern electrojet [128]. The results for additional reactive power losses for various storm scenario are presented in Fig. 5.12.

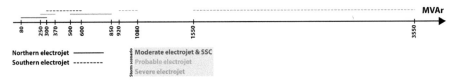

Figure 5.12: Calculated additional reactive power losses, MVAR, in National grid, UK, for various storm scenarios

Strategies of grid de-coupling by inserting series capacitors and the application of neutral blocking devices cannot be efficient without advanced warning (see Section 5.2). In addition to technical measures, operators can improve robustness to GMD by new operational routines, which are discussed further.

## 5.4   OPERATIONAL PROCEDURES

There is a certain belief that adopting operational procedures is the quickest way to make changes in addressing GMD threat. Operational procedures should be directed to protect power grid instability, uncontrolled separation or cascading failures of the bulk power system, caused by any damage to critical or vulnerable power system equipment. New constraints for power grid operation have appeared in the last years. Nowadays, the interconnected power grids are operated closer and closer to their limits and are becoming more and more interdependent. This leaves less operational margin when a GMD event impacts the grid. It implies a specific attention to the impact of cross-border disturbances on the whole interconnected power grid operation. The legal norms, which govern the power grid stability aspects, focus on:

1. Maintain acceptable voltage levels.

2. Maintain operation within stability limits.
3. Maintain operation within transfer limits.
4. Minimize the risk of cascading interruptions to the transmission systems.
5. Prevent physical damage to power system transmission facilities.
6. Eliminate thermal overloads.

The comparison of dispatching principles in ENTSOE, UPS of Russia and PJM is done together with comparative analysis of evolved practices for GMDs impact mitigation. The standards of the quality of power supply are nearly the same in aforementioned grids, but the operation principles, which ensure their fulfillment, are different. The differences are basically originated from architecture and the management methods of the power grids.

The aim of defining operating limits is to ensure the stable power system operation within the technical limits. The set of parameters that characterize the power system state is described in Section 4.1. The number of the data subsets ($U$, $I$, $P$, etc.) is related to the different power system situations, which are classified in relation to the probability of the emergency situation and urgency of the control actions. In ENTSOE power grid, the $N - 1$ principle requires the determination of acceptable voltage levels according to different scenarios of real-time simulation of power grid behavior in case of one network element loss. Four power system states are taken into account. Normal state confirms to the condition "No risk for system operation". If the operation within acceptable limits, but the risk for some operations exists, then the state is the alert one. In emergency state, the security principles are not full filled. The characteristics of blackout are: almost total absence of voltage in the transmission power grid; consequences abroad the dispatching zone of TSO. The acceptable voltage levels have to be estimated for the **N** situations and potentially different ones corresponding to **N − 1** scenario.

The architecture of UPS of Russia is represented by long transmission lines and weak interconnections. Under these circumstances, the satisfaction of the $N - 1$ principle cannot be provided. The approach using stability coefficients for power Kp and voltage $K_U$ is under consideration, which is described further.

If the network elements are loaded up to their thermal and stability limits, any contingency can violate the normal operation of the system. Thereby, the maximum power flow over each specified interface is identified. Interface consists of a number of high voltage facilities that divide power grids into two separate parts. Both transmission line and power transformer can form the interface. The following method for maximum power flow calculation is used.

1. Definition of the maximum power flow $P_m$ over the interface using different loading trajectories.
2. Choice of the minimum $P_m$ among the defined ones.
3. Calculation of the irregular oscillations of the power flow over the interface $\Delta P_{io}$, which depends on the load on the both sides of the interface $P_{l_1}$, $P_{l_2}$ (Eq. 5.5).

$$\Delta P_{io} = K \sqrt{\frac{P_{l_1} P_{l_2}}{P_{l_1} + P_{l_2}}}, \tag{5.5}$$

where $K$ is the coefficient, which is equal to 0.75 in case of automatic control of the power flow over the interface and is equal to 1.5 in case of manual control.

4. Calculation of the maximum admissible $P_{ma}$ (Eq. 5.6) and emergency admissible $P_{ea}$ (Eq. 5.7) power flows over the interface

$$P_{ma} = k_p^N P_m - \Delta P_{io} = 0.8 P_m - \Delta P_{io} \tag{5.6}$$

$$P_{ea} = k_p^e P_m - \Delta P_{io} = 0.92 P_m - \Delta P_{io} \tag{5.7}$$

Power stability coefficients for normal $k_p^N$ and emergency $k_p^e$ power system states are defined in Table 5.4 together with voltage stability coefficients $k_U$. The method is shown in Fig. 5.13. The value of maximum load $P_{max,load}$ corresponds to the amount of load $P_{l_2}$, which can be ensured without the loss of power system stability.

---

**Table 5.4**
**Stability coefficients**

| Power grid state | Stability coefficients | |
|---|---|---|
| | Power coefficient Kp | Voltage coefficient $k_u$ |
| Normal power grid state | 0.2 | 0.15 |
| Emergency power grid state | 0.08 | 0.1 |

---

The critical voltage level is defined as $0.7 \cdot U_{nom}$ for the nodes with the voltage level 110 kV and higher. The lower voltage limit for the normal power system state is $0.82 \cdot U_{nom}$ and for emergency state $0.78 \cdot U_{nom}$. These voltage levels must include a margin from the critical voltage. They can be the same for all network nodes or specific for particular ones due to stability requirements. The maximum admissible $P_{ma}$ and emergency admissible $P_{ea}$ power flows should be reduced in cases when voltage stability is violated.

PJM operates the grid considering voltage and stability related transmission limits as follows:

1. **Voltage Limits**: High, Low (NL), and Load Dump (LDL) actual voltage limits, High and Low emergency voltage limits (EL) for contingency simulation, and voltage drop limits for wide area transfer simulations to protect against wide area voltage collapse.

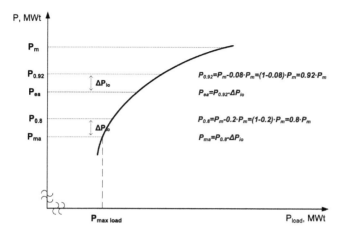

Figure 5.13: Calculation of the admissible power flow over the interface

2. **Transfer Limits**: The MW flow limitation across an interface to protect the system from large voltage drops or collapse caused by any viable contingency.
3. **Stability Limits**: Limit based on voltage phase angle difference to protect portions of the PJM RTO from separation or unstable operation.

PJM specifies admissible voltage limits as a function of operating voltage level (Table 5.5). The operator determines a collapse point for each interface. Admissible power transfer is limited by the voltage stability.

**Table 5.5**

**Admissible voltage deviations in PJM power grid**

| Limit, p.u. | 230 kV | 345 kV | 500 kV | 765 kV |
|---|---|---|---|---|
| 1.00 | - | - | NL | - |
| 0.97 | - | - | EL | - |
| 0.95 | NL | NL | LDL | NL |
| 0.92 | EL | EL | - | EL |
| 0.90 | LDL | LDL | - | LDL |

Since GMD impact on power grid may result in voltage avalanche (see Section 4.2), the accurate modeling of load should be done. Normally the load is represented as constant impedance or as a constant power take-off. For better representation of load node behavior, the static load characteristics are used in UPS of Russia. Static load characteristic is the sum of the corresponding characteristics of the loads connected to the node (the network power losses are also taken into account). Aggregated characteristic is defined mainly by the asynchronous electrical machines, transformer saturation and FACTS devices. The static load characteristics are described as in Eqs. 5.8 –5.9.

$$P_{nom} = P_{nom_0} \left( a_0 + a_1 \frac{U}{U_{nom}} + a_2 \left( \frac{U}{U_{nom}} \right)^2 \right) \tag{5.8}$$

$$Q_{nom} = Q_{nom_0} \left( b_0 + b_1 \frac{U}{U_{nom}} + b_2 \left( \frac{U}{U_{nom}} \right)^2 \right) \tag{5.9}$$

Parameters $P_{nom_0}$ and $Q_{nom_0}$ are the forecasted levels of active and reactive power consumption. Coefficients $a_0$, $a_1$, $a_2$, $b_0$, $b_1$ and $b_2$ are polynomial coefficients defined experimentally for each network node. The conditions $P = P_{nom_0}$ and $Q = Q_{nom_0}$ in case of $U = U_{nom}$ are held on the assumption of $a_0 + a_1 + a_2 = 1$ and $b_0 + b_1 + b_2 = 1$. The typical load characteristic for active power is $a_0 = 0.83$, $a_1 = -0.3$ and $a_2 = 0.47$. The compensation level of reactive power significantly influences the value of the coefficients $b_0$, $b_1$, $b_2$. In order to minimize this influence, the integrated load characteristic is used, which is based on the power factor $\cos\phi$. In this case, the typical static load characteristics for the voltage levels 6–35 kV are:

| | | | |
|---|---|---|---|
| $0.815 \leq U/U_n \leq 1.2$ | 4.9 | $-10.1$ | 6.2 |
| $U/U_n \leq 0.815$ | 0.657 | 0.343 | 0 |
| $U/U_n \geq 1.2$ | 1.708 | 0 | 0 |

For the voltage levels 110–220 kV the polynomial coefficients are:

| | | | |
|---|---|---|---|
| $0.815 \leq U/U_n \leq 1.2$ | 3.7 | $-7$ | 4.3 |
| $U/U_n \leq 0.815$ | 0.721 | 0.279 | 0 |
| $U/U_n \geq 1.2$ | 1.49 | 0 | 0 |

In the common practice of ENTSOE and PJM, stability limits are calculated without considering static load characteristics, since these power grids are characterized by the absence of power generation deficit and the presence of relatively short transmission lines. Authors recommend the adaptation of static load characteristics for stability transfer identification during severe GMDs in the practice of other TSOs.

If minimum voltage levels are identified in correlation to stability limits, high voltage limits are equipment related. Too high voltage can accelerate equipment aging and isolation degradation. The identified upper voltage limits are comparable in all three studied grids. The coefficients given by PJM are listed below. The possible overload for 765 kV network is 1.05; for 500 kV network – 1.1; for 345 kV network – 1.05; for 235 kV network – 1.05. Another equipment-related parameter is maximum current flow over network element. Seasonal and weather conditions are taken into account while defining current flow limits. In addition to them, the isolation type and service period are also taken into account in UPS of Russia. In $N-1$ or $N-k$ situations the overload protection implements tripping of the network element, if the loading has not come back under a given value. The multistep devices are used. The first step sends a signal, and the last one trips the network element. The intermediate ones puts into operation the control actions, which reduce the generation in the redundant part or the load in the deficit part of the system.

Power grid operation is exposed to the variety of disturbances that result in different ways on the power system behavior. The disturbances that constitute a menace to the power grid stability correspond to the reasonable reliability level. They are considered as a normative one. A contingency is considered as the trip of an element that cannot be predicted in advance. A scheduled outage is not a contingency. An "old lasting" contingency is considered as a scheduled outage.

In general, all the contingencies are subdivided into three groups: normal (normative), exceptional and out of range. The list of exceptional contingencies is defined by each TSO on its own risk evaluation in addition to the normal contingencies list. The exceptional contingencies list complies with the principles "no cascading impact outside my borders". The out of range contingencies are not considered.

All normative contingencies are subdivided in three groups (I, II, III) as a function of their severity and voltage level. The reliable operation of UPS of Russia has to be provided in case of disturbance of level III occurrence. Meanwhile, the power grid stability has to be ensured in case of disturbance of level II occurrence in the service state (maintenance of network element). The contingencies list is represented in Table 5.6. The ones, which are marked in italic, are considered as exceptional ones in ENTSOE and PJM.

In addition, PJM models system ability to withstand disturbances beyond those, which are reasonably expected. The list of these less probable contingencies is given below:

1. Sudden loss of the entire generation capability of any station for any reason.
2. Sudden loss of all lines on a single right-of-way.
3. Sudden dropping of a large load or major load center.
4. Sudden loss of all lines and transformers of one voltage emanating from a substation or switching station.
5. Failure of a fully redundant special protection system to operate when required.
6. Operation, partial operation or maloperation of a fully redundant special protection system for an event of condition for which it was not intended to operate.

The loss of one power transformer unit due to its core saturation by GIC corresponds to the cases "shutdown of the network element using the primary protection as a result of the three-phase fault with unsuccessful reclosure" and "Shutdown of the network element using redundant switch operator as a result of the three-phase fault with the fault of the one breaker". These contingencies are equivalent to the contingencies of III type for the network 750 kV. The loss of the power transformer in the "one and a half breaker" busbar scheme agrees with the contingency of group III. Nevertheless, current legal norms do not take into account the sudden loss of $N-2$ units.

## Table 5.6
## List of normative contingencies

| Contingency | Groups of the contingencies | | | |
|---|---|---|---|---|
| | 110– 220 kV | 330– 500 kV | 750 kV | 1150 kV |
| **Short Circuit on the network element (except busbar)** | | | | |
| Shutdown of the network element using the primary protection as a result of the single-phase fault (with the unsuccessful re-closure) | I | II | I | II |
| Shutdown of the network element using the backup protection as a result of the single-phase fault with the unsuccessful re-closure | II | II | III | III |
| Shutdown of the network element using the primary protection as a result of the three-phase fault with the unsuccessful re-closure | II | - | - | - |
| Shutdown of the network element using the primary protection as a result of the double-phase-to-ground fault (with the unsuccessful re-closure) | - | II | III | III |
| Shutdown of the network element using redundant switch operator as a result of the single-phase fault with the fault of the one breaker | III | III | III | III |
| *N − 1 double circuit line in the same corridor over a distance, which is more than a half of the length of the shorter line, as a result of the contingency of the group I* | III | III | III | III |
| *Shutdown of the network element as a result of the contingencies of the groups I and II in the case when the remedial of the breaker occurs the shutdown of the network element, which is connected to the same busbar* | III | III | III | III |
| **Short Circuit on the busbar** | | | | |
| Shutdown of the busbar as the result of the single-phase fault with disconnection | III | II | II | II |
| **Emergency power imbalance** | | | | |
| Shutdown of the: | II | II | II | II |
| ▪ generator | | | | |
| ▪ bloc of the generators at gas power plant | | | | |
| ▪ bloc of the generators connected to the one reactor at nuclear power plant | | | | |
| ▪ active power drop for at least 10 minutes at wind power plant | | | | |
| ▪ active power drop for at least 10 minutes at solar power plant | | | | |

Once a contingency occurs the system is readjusted as required using remedial actions. Two types of remedial actions are used: curative and preventive. Curative remedial actions, which have to be defined in advance and their efficiency has to be proven by simulation, are implemented after a disturbance in order to quickly relieve violated constraints of the system. Preventive remedial actions are decided and implemented in advance in cases when curative remedial actions could be not efficient or do not exist. The generalized characteristics of remedial actions used in ENTSO-E, PJM and UPS of Russia practices are given in Table 5.7. The labels are as following: **P** states for preventive remedial action; **C** states for curative remedial actions; and **A** shows that the remedial action is implemented automatically; **M** means manually.

**Table 5.7**
**Characteristics of remedial actions**

| Remedial action | Time constant | P/C | A/M | Controlled parameter |
|---|---|---|---|---|
| Topology change | A few seconds | P/C | A/M | power flow, voltage |
| Usage of phase shifter transformer | A few seconds | P/C | A/M | power flow, voltage |
| Reduction of interconnection capacities | A few minutes | P/C | M | power flow, voltage |
| Start-up of tertiary reserve | Depends on generation unit type | P/C | A/M | power flow, voltage |
| Generation unit shut-down | A few seconds | C | A/M | power flow, voltage |
| Load shedding | A few milliseconds | P/C | A/M | power flow, voltage |
| Limiting of intraday trade | A few minutes | P | M | power flow, voltage |
| Switching on/off shunt reactors, capacitors | A few minutes | P | M | power flow, voltage |

All kinds of remedial actions have a total cost of their implementation. The principle of quality power supply requires, firstly, the implementation of the remedial actions of the lowest cost. Load shedding as an action, associated with the energy undersupply, is not considered as a primary one. It is proposed as a guide that the amount of 50% of total load can be operated under load shedding. This amount of load is a maximum one which can be achieved without leading to the instability occurrence. It seems to be a common practice in many systems worldwide, except the US, where 30% limit is applied since improvements after the blackout in 2003. The probability of full volume activation for load shedding is low; however, this amount should be available during the operation. Corresponding measurements are done continuously.

GMDs are considered by PJM as the reason to operate grid more conservatively. If NOAA issues a warning or an alert messages of potential GMD of severity Kp=7 or greater, PJM implements the following operation plan. When GIC measurements exceed the associated GIC operating limit at one and only one of the transformers, PJM may take an action as soon as necessary, but must take an action if conditions persist for 10 minutes. Generation dispatchers provide as much advance notification as possible regarding details of more restrictive plant procedures that may result in plant reductions to protect equipment. Upon identification of a GMD, PJM declares GMD action and operates the system to GMD transfer limits that are determined from modeling different scenario. PJM actions include:

1. PJM Dispatch notifies members and neighbors of a GMD action to mitigate the effects of GMD on the system. PJM begins to operate the system to the geomagnetic disturbance transfer limits.
2. When the GMD transfer limit is approached or exceeded, generation re-dispatch assignments are made in the most affected areas to control this limit. PJM Dispatch also evaluates the impact of the existing inter-area transfers and modifies the schedules that adversely affect the GMD transfer limit. If insufficient generation is available to control this limit, the emergency procedures are implemented. Dispatcher has to validate the GMD transfer limit and develop a voltage drop curve for the GMD transfer limit contingency. Pre-contingency load shedding is not used to control transfers to the GMD transfer limit.
3. After the GIC measurements at all monitored transformers have fallen below the associated operating limit, PJM Dispatch continues to operate the system to the GMD transfer limit for a period of 3 hours. If the measurement values are confirmed to remain below the GIC limits, PJM Dispatch cancels the GMD transfer limit.

After Hydro-Québec event in 1989, the local operator adapted in its practice also pre-caution actions whenever any critical system asymmetry caused by a GMD is observed. The precautions include: (1) spreading the generation more evenly; (2) increasing the spinning reserve; (3) putting back in service as many lines in the system as possible; (4) reducing transmission on the lines to about 80/90% of their rated capacity; (5) suspending any tests under way on the system; (6) reducing the number of switching operations on the system to a minimum to avoid causing instability; (7) modifying, or remotely modifying, the static-compensator tripping signals to alarm mode only, thereby forcing the compensators to remain in service [161].

It was shown before that GMDs cause the increase of reactive power deficit, voltage drops and disruption of interconnected system operations. The remedial action "network split" is widely use in the practice of UPS of Russia for preventing small signal instability. The adaptation of this remedial action is recommended for other TSOs. The command for "network split" has to be given after receiving the information about measured GIC in the power transformers. The analysis of blackouts caused by GMDs showed that $N-1$ principle is not sufficient. Authors propose to consider $N-2$ principle, especially in the immensely growing power grids of megapolis.

Depending on the level of the disturbance and available resources, TSO may implement "bottom-up", "top-down" or combination of these two approaches to restore the system. There is little operational data on the effectiveness of currently available technology solutions to mitigate the effects of a large-scale GMD [169]. Overall, operational control algorithm as a measure is reactive at best and cannot provide optimal GIC negative effects mitigation.

## 5.5 LEGISLATIVE PROCEDURES

Stakeholders frequently have different interpretation of the risk which complicates the process [133]. The goal of legislative procedures is to attribute the risk that can be mitigated by aforementioned generic capabilities. In general, the process includes following elements: (a) measures to raise awareness of the hazard, and the consequent risks, across national, regional, and local government; (b) engagement with industry and academia to better understand the risks and the options for mitigation (including research to improve understanding and mitigation); (c) establishment of high-level plans to deal with the risk, most importantly including the capability to adapt plans to the actual impacts (reflecting that risks are extremely unlikely to occur in the exact same form as used in planning scenarios); (d) identification and development of capabilities to assist in mitigation of the risk, including both capabilities such as forecasting to reduce risk impacts and capabilities such as equipment and trained staff to speed recovery following the risk impact [134].

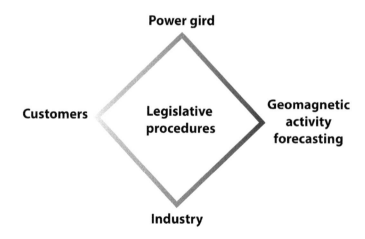

Figure 5.14: Mitigation square

GMD impact on power grids is a complex and technical issue but, many of the potential consequences are common to other risks. Plans to respond to severe GMDs can be dealt under existing plans for other events. Mitigation design and planning for GMD impact minimization requires the full understanding of the square (Fig. 5.14) and work alongside other stakeholders. The process by which the practical problems are simplified into a decision aid model could be subjective [135]. The comprehensive evaluation of a square requires a full-scale analysis of: geomagnetic activity, power system operation principles, customers response to geomagnetic activity, industry response to geomagnetic activity. Interdisciplinary approaches already proved to be beneficial for extreme event mitigation.

Several countries have recognized the risks associated with severe GMDs and have launched initiatives to prevent, prepare for and respond to the threat. They specified the tasks that will lead to improvements in policies, practices and procedures for decreasing the vulnerability. It became an important issue for the policy community, especially, since modern economies are highly depended on reliable electricity supply.

The first legal norm regarding GMD impact on power grids was FERC Order no. 779, which went into effect in 2014. It asks a utility to have an operational plan to mitigate the impact of GMD on their system [116]. NERC conducts annual audits of various electric utilities over the course of the year, either randomly or after a major outage [156]. FERC then imposes fines on utilities for violating NERC reliability standards. The white paper on penalty pertaining to violations of critical infrastructure protection reliability standards gives a good indication of procedures between FERC and NERC [136].

In 2014, the White House integrates efforts to develop a coordinated federal government response to space weather and established the Space Weather Operations, Research and Mitigation (SWORM) Task Force, co-chaired by the White House Office of Science and Technology Policy, the Department of Commerce [137]. The latest high-level attention is centered around regulatory actions initiated by Federal Energy Regulatory Commission, US. The National Space Weather Strategy is released concurrently with National Space Weather Action Plan in 2015. The six national strategic goals for leveraging existing policies and ongoing research and development efforts are specified. These goals are:

1. Establish benchmarks for space weather events – sets of physical characteristics and conditions against which a space weather event can be measured. The benchmark addresses the different types of space weather activity: induced geo-electric fields, ionizing radiation, ionospheric disturbances, solar radio bursts, upper atmospheric expansion. The induced geo-electric field benchmark will define the amplitude of the induced $E$-field and time dependence of the induced $E$-field. At a minimum, the ionizing radiation benchmark defines the radiation intensity as a function of time, particle type and energy. The ionospheric disturbance benchmark defines the ionospheric radio absorption, total electron content, ionospheric refractive index, peak ionospheric densities and the height of the peak. The benchmark for solar

radio bursts define the wavelength and frequency bands of the relevant solar radio bursts and flux in these bands. The upper-atmospheric expansion benchmark defines neutral density, winds, composition and the temperature of the thermosphere. The benchmarks are developed for the event with occurrence rate 1-in-100 years and for the intensity level of the theoretical maximum of the event. The benchmarks focus on the modeling of the event not on the impacts. Further goals target the impact modeling.

2. Enhance response and recovery capabilities – an improved ability to forecast and understand the effects and the magnitude of a GMD event. The objectives of the second goal include: complete an all-hazard outage response and recovery plan; support government and private sector planning for and management of extreme GMD events; provide guidance on contingency planning for the effects of extreme GMD for essential government and industry services; ensure capability and interoperability of communication systems during extreme GMD events; encourage owners and operators of infrastructure and technology assets to coordinate development of realistic power-restoration priorities and expectations.

3. Improve protection and mitigation efforts – development of capabilities and actions to secure the Nation from the effects of GMD, including vulnerability reduction. Mitigation focuses on minimizing risks, addressing cascading effects and enhancing resiliency to disasters. The objectives of the third goal include: encourage development of hazard-mitigation plans that reduces vulnerabilities to, manage risks from, assist with response to the effects of GMD; work with industry to achieve long-term reduction of vulnerability to GMD events by implementing measures at locations most susceptible to GMD.

4. Improve assessment, modeling and prediction of impacts on critical infrastructure – the understanding of vulnerabilities, growth of situational awareness and development of capabilities to predict impacts on all affected critical infrastructure systems. The objectives of the fourth goal include: assess the vulnerability of critical infrastructure systems to GMD; develop a real-time infrastructure assessment and responding capability; develop and refine operational models that forecast the effects of GMD on critical infrastructure; improve operational impact forecasting and communications; conduct research on the effects of GMD on industries, operational environments and infrastructure sectors.

5. Improve GMD services through advancing understanding and forecasting – develop timely and accurate space weather information products for ensuring that emergency managers, first responders, government officials, business and the public will be empowered to make fast, smart decisions in response to space weather events. The objectives of the fifth goal include: improve understanding of users needs for space weather forecasting to establish lead-time and accuracy goals; ensure that space-weather products are intelligible and actionable to inform decision making; establish and sustain

a baseline observational capability for space-weather operations; improve forecasting lead-time and accuracy; enhance fundamental understanding of space weather and its drivers to develop and continually improve predictive models; improve effectiveness and timelines of the process as transitions research to operations (R2O).

6. Increase international cooperation – exchange of best practices between the US and international partners strengthen global capacity to respond to extreme GMD events. The objectives of the sixth goal include: build international support and policies for acknowledging space weather as a global challenge; increase engagement with the international community on observation infrastructure, data sharing, numerical modeling and scientific research; strengthen international coordination and cooperation on space weather products and services; promote a collaborative and cooperation on space weather products and services; promote a collaborative international approach to preparedness for extreme space weather.

The National Space Weather Action Plan defines approximately 100 action items and correlated timelines. Later, the enacted 2015 Fixing America's Surface Transportation Act Public Law (114-94) identifies GMD as a potential cause of power grid security emergency and amends the Federal Power Act (16 United States Code Section 824 et seq.), giving the Secretary of Energy, upon the declaration of an emergency by the President, the authority to issue orders to protect the reliability of critical electric infrastructure (16 United States Code Section 824 (O-1)(b)(1)) [138]. In response to the action plan, US Department of State and Secure World Foundation hold a panel on GMD as a global challenge to push further the international collaboration on this topic.

The risk of severe GMD was added to the National risk assessment in 2011 and the subsequent national risk was registered in 2012 in UK. The UK approach to GMD preparedness is underpinned by three elements:

- designing mitigation into infrastructure where possible;
- developing the ability to provide alerts and warnings of space weather and its potential impacts;
- having in place plans to respond severe events.

According to UK Cabinet Office, the main challenge is the low risk awareness level. Therefore, one of the main priorities of the future work is the development of international collaboration to fill gaps in space and ground capabilities to monitor, predict and assess the events. The other main difficulties are: the difficulty of accurate event forecasting; the short warning time to prepare once we have uncertainty about the size of the event; understanding potential impact in the modern developed societal and technological infrastructure; a lack of monitoring capability of the events once they start. Nevertheless, considerable progress has been already achieved. Royal Academy of Engineering published a report on severe space weather in February 2013. Additionally, the COBR response guide for severe space weather was produced in 2013 and updated in 2015. Met Office launched forecasting capability in 2014 and coordinated warnings and alerts transfer to the National Grid. In order to communicate the message that GMD risk has to be shared by everyone, UK Space Weather Public Dialogue was established within the members of the public that interact with scientists or other stakeholders [139].

Though there is no pan-European in dealing with extreme GMD threat, six European countries (Finland, Hungary, Netherlands, Sweden, UK, Norway) have included space weather in their risk portfolio, National Risk Assessment (NRA). In the Swedish NRA, GMDs are identified as a particular serious threat and possible impact scenario is presented. This scenario assumed the occurrence of the Carrington-type event with major disruptions to satellite signals and HF communications, as well as power blackout in southern and central Sweden. It is aimed to understand the potential consequences for over the 8 million people living in the affected area (total population of Sweden is ca. 10 million people). Overall evaluation of the current legal framework shows ineffectiveness in regional approach in dealing with the threat due to increased interconnected nature of modern systems.

Efforts were taken to promote the risk of severe GMDs problem for mid- and low-latitude regions. Nowadays, TSOs in nations such as the US, UK, Canada, Finland, Norway, Sweden, China, Japan, Brazil, Namibia, South Africa and Australia have launched hazard assessment programs to understand and mitigate the possible GIC impact on their power grids [140].

The list of countries that consider critical infrastructure failure due to power loss is much wider. Twenty-four EU Member States discuss the primary scenario of critical infrastructure destruction in addition to the loss of critical infrastructure as a secondary effects of the hazards. Twenty EU Member States include scenarios of long-lasting power outages [141]. Severe GMDs can be considered as a possible trigger. On the Pan-European level the European Commission has proposed a regulation on risk-preparedness for the prevention, preparation for and management of electricity crisis situations [142]. Within this regulation, ENTSO-E is delegated with the task to develop a methodology and to identify electricity crisis scenario every three years or sooner if needed. The assessment should cover rare and extreme natural hazards, simultaneous accidental hazards, consequential hazards including fuel shortages, malicious attacks. The risks should be ranked according to impact severity and appearance probability. Based on the assessment results, each Member State should establish a risk-preparedness plan, which consists of two parts: local (national) actions or regional (cross-border) actions. It is also stressed that all the plans should follow the principle "market first", i.e. markets should be permitted to operate under scarcity conditions, where electricity prices spike. The existing assessment procedure includes seasonal outlooks preparation according to the Article 8 of the Electricity Register, which are mainly focused in their current form on the generation adequacy. Article 4 of the Electricity Directive and Article 7 of the Security of Supply (SoS) Directive impose the general obligation to Member States to monitor security of supply and to publish bi-annual report outlooking their findings.

The policy recommendations to mitigate the space wether negative effects are summarized in [143]. The first five recommendations are focused on disaster management, while others focus on improving power grid resiliency.

- **Recommendation 1**: The importance of using a consistent set of realistic scenario is stressed. This measure helps to share a common understanding of a hazard and corresponding threats. Currently, space weather disaster scenarios with the three return periods are studied: 1-in-10 years, 1-in-30 years and 1-in-100 years.
- **Recommendation 2**: It is suggested to consider events with recurrence intervals longer than 100 years as well.
- **Recommendation 3**: The results achieved within previous step should be pushed further. Mitigation strategies based on a common scenario should be implemented across each policy area.
- **Recommendation 4**: The need for developing, implementation and practicing emergency operation plans before disaster strikes is emphasized. These emergency plans should describe emergency repair and recovery actions, assign responsibilities, identify resources and address coordination and communication [144]. All TSOs should annually participate in local, regional, national and international exercises and incorporate the findings. It is shown that the use of Outage Management System (OMS) helps to prioritize restoration of power as well as to dispatch, track and manage repair crews [145].
- **Recommendation 5**: The critical customers should be prioritized. The example of how critical customers can be described, and how this information can be integrated in the emergency management is given in Table 5.8 [146]. Further recommendations stress the power grid resiliency problem.
- **Recommendation 6** The transition from hardening system assets and facilities to building resilience into the grid is proposed. Traditional hazard mitigation strategies are focused on strengthening power grid components based on the assessed risk level. These measures may be prohibitively expensive or impractical [147]. Building power grid and resiliency instead would not focus on preventing damage from a single hazard, but rather to enable TSO to continue functioning when critical components are failed and swiftly return to operation after a disruption. Experience shows that the availability of spears and replacement parts and equipment for critical assets and facilities is crucial.
- **Recommendation 7** It is promoted to stockpile the repair or replacement of key assets and equipment. The example of such a program named STEP is given in Section 5.3

There is a need to engage stakeholder groups in a continuous dialogue and coordinate a provision of roles, perspectives and goals in the process of the development and implementation of safe guidelines and good practices consistent with recent scientific advances [148]. The public awareness of risks is limited by disaster

**Table 5.8**

**Example on how to prioritize critical customers and integrate this information in emergency planning**

| Parameter | $Event_1$ | $Event_i$ | $Event_n$ |
|---|---|---|---|
| **Customer** | | | |
| Designation | - | - | - |
| Type | - | - | - |
| Location | - | - | - |
| Outage consequences | - | - | - |
| Minimum power required | - | - | - |
| **Grid** | | | |
| TSO | - | - | - |
| Consequences | - | - | - |
| Capabilities | - | - | - |
| Needs | - | - | - |

prevention plans. The society is more aware about already experienced threats. Consequently, policy and strategies have to be design in align with the society capacity (Section 6.3.4). Better understanding of the likelihood and zones prone to GMDs, technology vulnerability and the operational performance of potential solutions will help decision-makers to determine the most optimal and appropriate for GMD effects mitigation solutions.

## 5.6  CONCLUSION

At the end of the International Decade for Natural Disaster Reduction (INDRR), the UN Secretary-General Kofi Annan insisted on a need for a "transition to a culture of prevention" [149]. Twenty years have passed since that statement was made. The UN's International Strategy for Disaster Reduction, the International Federation of Red Cross, the World's Bank Global Facility for Disaster Reduction and Recovery and the EU's Strategy for Supporting Disaster Risk Reduction are examples of the growing community of organizations committed to increasing resilience of states to natural hazards [150]. Intensive steps were taken to mitigate terrestrial hazards. GMD mitigation stays far behind more common and known threats. Currently, most organizations that are vulnerable to extreme GMD events do not have systematic emergency plans [141]. The most important role that science plays in legislative procedures is to sustain the interest of relevant stakeholders in GMD, to show that it is an event that needs attention.

The National Infrastructure Protection Plan proposed a framework for risk mitigation, which can be adopted for the GMD risk mitigation. It is not a perspective approach for risk management, rather, it is a construct for risk management. The framework is presented in Table 5.9.

**Table 5.9**

**Risk management framework**

| Phases | Related actions |
| --- | --- |
| Set goals & objectives | - Set of broad goals for infrastructure security and resilience<br>- Determine collective actions through joint planning efforts |
| Identify infrastructure | - Analyze associated dependencies and interdependencies. It was observed that not all actors are involved in the process from the same perspective [151] |
| Assess & analyze risk | - Improve information sharing<br>- Apply knowledge to enable risk-informed decision making |
| Implement risk management activities | - Rapidly identify, assess, and respond to cascading effects during and following incidents<br>- Promote infrastructure, community, and regional recovery following incidents |
| Measure effectiveness | - Learn and Adapt During and After Exercises and Incidents |

Dealing with the risk of a 100-year event is not a day-to-day business of grid owners and operators. However, there is an overarching societal interest in avoiding catastrophic GMD impacts, thus day-to-day challenges should not prevent operators in constructing resilient power grids. In general, all these mitigation actions can be the part of one of the risk strategies, which are chosen with respect to financial impact and likelihood of event occurrence. They can be classified as follows [152]:

1. **Accept the risk** – activities responding to a practice of monitoring the risk if cost-benefit analysis determines the cost to mitigate risk is higher than the cost to bear the risk. The best response in this case is to accept the risk.
2. **Transfer the risk** – activities with low probability of occurring, but with a large financial impact. The best response is to transfer a portion or all of the risk to a third party by purchasing insurance, hedging, outsourcing, or entering into partnerships.
3. **Mitigate the risk** – activities with a likelihood of occurring, but financial impact is small. The best response is to use management control systems to reduce the risk of potential loss.

4. **Avoid the risk** – activities with a high likelihood of loss and large financial impact. The best response is to avoid the activity.

In the view of GMD, two strategies can be chosen: **"transfer the risk"** or **"avoid the risk"**. The mitigation actions for increasing power grid resiliency to GMDs have to be specified on the contribution of each measure for improving resiliency. Thereby, the cost/benefit analysis of each action has to be performed. Following the analysis, the mitigation actions can be ranked and further implemented based on both their resilience and cost/benefit performance. The time after an event can be considered as a window of opportunities for implementing the actions. The relevant stakeholders in order to avoid the same loss in the future are eager to implement actions with higher cost and boost long-term resiliency. For example, the Hydro-Québec blackout in 1989 gave a kick to redesign the entire high voltage grid by installing lengthwise series compensation equipment.

Infrastructure hardening is the only pro-active mitigation measure. Such modifications to the power grid are more reliable than operational mitigation strategies, as they provide a guaranteed solution, compared with the former, which relies on the situational awareness of an operator to mitigate the problem. There should be a transition from hardening system components and facilities to building resilience into the power grid to enable the system to function even under disaster conditions or recover more quickly. The pathway to the strategy **"trabsfer the risk"** is described in Section 6.6.

## REFERENCES

1. Panteli, M., Trakas, D. N., Mancarella, P., Hatziargyriou, N. D. (2017). Power systems resilience assessment: Hardening and smart operational enhancement strategies. In *Proceedings of the IEEE*, 105(7), 1202–1213. doi: 10.1109/JPROC.2017.2691357.
2. Hokstod, P., Utne, I.B., Vatn, J. (2012). *Risk, and Interdependencies in Critical Infrastructures*. Springer series in Reliability Engineering. Springer-Verlag, London.
3. Leitch, M. (2010). ISO 31000: 2009-The new international standard on risk management. *Risk analysis*, 30(6), 887.
4. Ramalingam, B., Jones, H., Reba, T., Young, J. (2008). *Exploring the Science of Complexity: Ideas and Implications for Development and Humanitarian Efforts*, 285. Overseas Development Institute, London.
5. Snowden, D., Boone, M.J. (2007). A leader's framework for decision making: Wise executives tailor their approach to fit the complexity of the circumstances they face. *Harvard Business Review*, 85(11), 68.
6. De Rosa, J.K., Grisogono, A.M., Ryan, A.J., et al. (2008). Research agenda for the engineering of complex systems. *IEEE International Complex Systems Conference SysCon 2008 Montreal*.
7. Klein, R., Nicholls, R., Thomalla, F. (2003). Resilience to natural hazards: How useful is this concept?. *Global Environmental Change Part B: Environmental Hazards*, 5:35–45.
8. Holling, C.S. (1973). Resilience and stability of ecological systems. *Annual Review of Ecology and Systematics*, 4(1), 1–23.

9. United Nations (2009). *2009 UNISDR Terminology on Disaster Risk Reduction*. United Nations, Geneva.

10. Cabinet Office (2011). *Keeping the Country Running: Natural Hazards and Infrastructure*. Cabinet Office, London.

11. Cimellaro, G., Reinhyorn, A., Bruneau, M. (2010). Seismic resiliency of a hospital system. *Structure and Infrastructure Engineering*, 6:127–144.

12. Ehlen, M.A., Vugrin, E.D., Warren, D.E. (2010). A resilience assessment framework for infrastructure and economic systems: Quantitative and qualitative resilience analysis of petrochemical supply chains to a hurricane. *Process Safety Progress*, 30(3), 280–290.

13. Berkeley, A. R., Wallace, M., Coo, C. (2010). A framework for establishing critical infrastructure resilience goals. *Final Report and Recommendations by the Council, National Infrastructure Advisory Council*.

14. Gasser, P., Lustenberger, P., Cinelli, M., Kim, W., Spada, M., Burgherr, P., Sun, T. Y. (2019). A review on resilience assessment of energy systems. *Sustainable and Resilient Infrastructure*, 1–27. doi: 10.1080/23789689.2019.1610600.

15. Gasser, P., Suter, J., Cinelli, M., Spada, M., Burgherr, P., Hirschberg, S., Stojadinović, B. (2020). Comprehensive resilience assessment of electricity supply security for 140 countries. *Ecological Indicators*, 110, 105731. doi: 10.1016/j.ecolind.2019.105731

16. Nateghi, R., Reilly, A. C. (2017). All-hazard approaches to infrastructure risk reduction: Effective investments through pluralism. In *Safety and Reliability–Theory and Applications Briš 2017*. Taylor and Francis Group. CRC Press, London.

17. Cauley, G., Lauby, M. (2010). *High-Impact, Low-Frequency Event Risk to the North American Bulk Power System*. North American Electric Reliability Corporation.

18. Pescaroli, G., Wicks, R. T., Giacomello, G., Alexander, D. E. (2018). Increasing resilience to cascading events: The M. OR. D. OR. scenario. *Safety Science*, 110, 131–140. doi: 10.1016/j.ssci.2017.12.012.

19. Panteli, M., Mancarella, P. (2015). The grid: Stronger, bigger, smarter? Presenting a conceptual framework of power system resilience. *IEEE Power and Energy Magazine*, *13(3)*, 58–66.

20. Francis, R., Bekera, B. (2014). A metric and frameworks for resilience analysis of engineered and infrastructure systems. *Reliability Engineering & System Safety*, 121:90–103.

21. Ouyang, M., Duenas-Osorio, L. (2014). Multi-dimensional hurricane resilience assessment of electric power systems. *Structural Safety*, 48:15–24.

22. Henry, D., Ramirez-Marquez, J.E. (2012). Generic metrics and quantitative approaches for system resilience as a function of time. *Reliability Engineering & System Safety*, 99:114–122.

23. Shinozuka, M., Chang, S. E., Cheng, T. C., Feng, M., O'rourke, T. D., Saadeghvaziri, M. A., Shi, P. (2004). *Resilience of Integrated Power and Water Systems*, 65–86. Multidisciplinary Center for Earthquake Engineering Research.

24. Espinoza, S., Panteli, M., Mancarella, P., Rudnick, H. (2016). Multi-phase assessment and adaptation of power systems resilience to natural hazards. *Electric Power Systems Research*, 136:352–361.

25. Cavallo, A., Ireland, V. (2014). Preparing for complex interdependent risks: A system of systems approach to building disaster resilience. *International Journal of Disaster Risk Reduction*, 9:181–193.

26. Eismann, C. (2014). Trends in critical infrastructure protection in Germany. *Transactions of the VŠB–Technical University of Ostrava. Safety Engineering Series*, 9:26–31

27. Albertson, V.D., Bozoki, B., Feero, W.E., et al. (1993). Geomagnetic disturbance effect on power systems. *IEEE Transactions on Power Delivery*, 8:1206–1216.

28. Gombosi, T., Baker, D., Balogh, J., et al. (2017). Anthropogenic space weather. *Space Science Reviews*, 212:3–4.

29. Howard, R. (2006). A historical perspective on coronal mass ejections. *Solar Eruptions and Energetic Particles*, 212:7–13.

30. Speiser, J. (2017). 22 Years of Solar and Heliospheric Observatory. Published on AGU website on 30 November 2017. Accessible `https://fromtheprow.agu.org/ 22-years-solar-heliospheric-observatory/`.

31. Kilpua, E.K.J., Lugaz, N., Mays, M.-L., Temmer, M. (2019). Forecasting the structure and orientation of earthbound coronal mass ejections. *Space Weather*, 17(4), 498–526. doi: 10.1029/2018SW001944.

32. Lugaz, N., Farrugia, C. J., Smith, C. W., Paulson, K. (2015). Shocks inside CMEs: A survey of properties from 1997 to 2006. *Journal of Geophysical Research: Space Physics*, 120(4), 2409–2427. doi: 10.1002/2014JA020848.

33. Tsurutani, B. T., Gonzalez, W. D., Tang, F., Lee, Y. T. (1992). Great magnetic storms. *Geophysical Research Letters*, 19(1), 73–76. doi: 10.1029/91GL02783.

34. Mays, M. L., Thompson, B. J., Jian, L. K., Colaninno, R. C., Odstrcil, D., Möstl, C., MacNeice, P. J. (2015). Propagation of the 2014 January 7 CME and resulting geomagnetic non-event. *The Astrophysical Journal*, 812(2), 145. doi: 10.1088/0004-637X/812/2/145.

35. Schwenn, R., Dal Lago, A., Huttunen, E., Gonzalez, W.D. (2005). The association of coronal mass ejections with their effects near the Earth. *Annales Geophysicae*. doi: 10.5194/angeo-23-1033-2005.

36. Lugaz, N., Farrugia, C. J., Huang, C. L., Spence, H. E. (2015). Extreme geomagnetic disturbances due to shocks within CMEs. *Geophysical Research Letters*, 42(12), 4694–4701. doi: 10.1002/2015GL064530.

37. Harrison, R., Davies, J., Rae, J. (2019). From heliophysics to space weather forecasts. *Astronomy & Geophysics, 60(5)*, 5–26. doi: 10.1093/astrogeo/atz178

38. Koza, F. (2011). August 2011 Coronal Mass Ejections PJM GIC Detector Data. *NERC Geomagnetic Disturbance Task Force Meeting August, 30-31, 2011*.

39. Sun, R., McVey, M., Lamb, M., Gardner, R.M. (2015). Mitigating geomagnetic disturbances: A summary of Dominion Virginia Power's efforts. *IEEE Electrification Magazine*, 3:34–45.

40. Posner, A., Hesse, M., Cyr, O. S. (2014). The main pillar: Assessment of space weather observational asset performance supporting nowcasting, forecasting, and research to operations. *Space Weather*, 12(4), 257–276. doi: 10.1002/2013SW001007.

41. Evans, D.L., Balch, B., Murtagh, D., et al. (2004). *Service assessment: intense space weather storms October 19 - November 7, 2003*. US Department of Commerce, National Oceanic and Atmospheric Administration, National Weather Service. Silver, Springer.

42. Schrijver, C. J., Rabanal, J. P. (2013). A survey of customers of space weather information. *Space Weather*, 11(9), 529–541.

43. Teisberg, T.J., Weiher, R.F. (2000). Valuation of geomagnetic storm forecasts: An estimate of the net economic benefits of a satellite warning system. *Journal of Policy Analysis and Management*, 19:329–334.

44. European Space Agency (ESA) (2016). *SWE Cost-Benefit Analysis*. Springer series in Reliability Engineering. Springer-Verlag, Paris.

45. Oughton, E. J., Hapgood, M., Richardson, G. S., Beggan, C. D., Thomson, et al. (2019). A risk assessment framework for the socioeconomic impacts of electricity transmission infrastructure failure due to space weather: An application to the United Kingdom. *Risk Analysis*, 39(5), 1022–1043. doi: 10.1111/risa.13229.

46. Thomas, S.-R., Fazakerley, A., Wicks, R.-T., Green, L. (2018). Evaluating the skill of forecasts of the near-earth solar wind using a space weather monitor at L5. *Space Weather*, 16(7), 814–828. doi: 10.1029/2018SW001821.

47. Shinozuka, M., Chang, S.E., Cheng, T.C., et al. (2004). Navigation analysis for an L5 mission in the Sun-Earth system. *AAS/AIAA Astrodynamicist Specialist Conference, Girdwood*. doi: 10.13140/2.1.2069.4402.

48. Gopalswamy, N., Davila, J.M., Cyr, O.C., et al. (2011). Earth-Affecting Solar Causes Observatory (EASCO): a potential international living with a star mission from Sun–Earth L5. *Journal of Atmospheric and Solar-Terrestrial Physics*, 73:658–663.

49. Trichas, M., Gibbs, M., Harrison, R., et al. (2015). Carrington-L5: The UK/US operational space weather monitoring mission. *Hipparchos*, 2:25–31.

50. Lavraud, B., Liu, Y. D., Harrison, R. A., Liu, W., et al. (2014). Instant: An innovative L5 small mission concept for coordinated science with Solar Orbiter and Solar Probe Plus. *AGUFM, 2014*, SH21B–4109.

51. Turner, D.L., Li, X. (2011). Using spacecraft measurements ahead of Earth in the Parker spiral to improve terrestrial space weather forecasts. *Space Weather*, 9(1). doi: 10.1029/2010SW000627.

52. Savani, N.P., Vourlidas, A., Szabo, A., et al. (2015). Predicting the magnetic vectors within coronal mass ejections arriving at Earth: 1. Initial architecture. *Space Weather*, 13:374–385.

53. Savani, N.P., Vourlidas, A., Richardson, I.G., et al. (2017). Predicting the magnetic vectors within coronal mass ejections arriving at Earth: 2. Geomagnetic response. *Space Weather*, 15(2). doi: 10.1002/2016SW001458.

54. Kay, C., Gopalswamy, N., Reinard, A., Opher, M. (2017). Predicting the magnetic field of Earth-impacting CMEs. *The Astrophysical Journal*, 835(2), 117. doi: 10.3847/1538-4357/835/2/117.

55. Patsourakos, S., Georgoulis, M.K. (2017). A helicity-based method to infer the CME magnetic field magnitude in Sun and geospace: Generalization and extension to Sun-like and M-dwarf stars and implications for exoplanet habitability. *Solar Physical*, 292(7), 89. doi: 10.1007/s11207-017-1124-1.

56. Baker, D.N., McPherron, R.L., Cayton, T.E., Klebesadel, R.W. (1999). Linear prediction filter analysis of relativistic electron properties at 6.6 RE. *Journal of Geophysical Research: Space Physics*, 95:15133–15140.

57. McPherron, R.L. (1998). Determination of linear filters for predicting $A_P$ during Jan. 1997. *Geophysical Research Letters*, 25:3035–3038.

58. Papitashvili, V.O., Clauer, C.R., Killeen, T.L., et al. (1998). Linear modeling of ionospheric electrodynamics from the IMF and solar wind data: Application for space weather forecast. *Advances in Space Research*, 22:113–116.

59. O'Brien, T.P., Sornette, D., McPherron, R.L. (2001). Statistical asynchronous regression: Determining the relationship between two quantities that are not measured simultaneously. *Journal of Geophysical Research: Space Physics*, 106:13247–13259.

60. Lundstedt, H. (1996). Solar origin of geomagnetic storms and prediction of storms with the use of neural networks. *Surveys in Geophysics*, 17:561–573.

61. Tulunay, Y., Sibeck, D.G., Senalp, E.T., Tulunay, E. (2005). Forecasting magnetopause crossing locations by using neural networks. *Advances in Space Research*, 36:2378–2383.

62. Wang, H., Ridley, A.J., Lühr, H. (2008). Validation of the space weather modeling framework using observations from CHAMP and DMSP. *Space Weather*, 6:1–16.

63. Valach, F., Revallo, M., Bochníček, J., Hejda, P. (2009). Solar energetic particle flux enhancement as a predictor of geomagnetic activity in a neural network-based model. *Space Weather*, 7(4), S04004.

64. Sharma, N., Sharma, P., Irwin, D., Shenoy, P. (2011). Predicting solar generation from weather forecasts using machine learning. *2011 IEEE International Conference on Smart Grid Communications (SmartGridComm)*,528–533.

65. Camporeale, E., Wing, S., Johnson, J. (2018). *Machine Learning Techniques for Space Weather*. Elsevier, Amsterdam.

66. Gombosi, T.I., Chen, Y., Manchester, W., et al. (2018). Machine learning and the" Holy Grail" of space weather forecasting. *American Geophysical Union Fall Meeting (AGUFM), 2018*, SM54A-02.

67. Lundstedt, H. (1997). AI techniques in geomagnetic storm forecasting. *Geophysical Monograph Series (GMS)*, 98:243–252.

68. Monfared, M., Rastegar, H., Kojabadi, H.M. (2009). A new strategy for wind speed forecasting using artificial intelligent methods. *Renewable Energy*, 34:845–848.

69. Codrescu, M.V., Fuller-Rowell, T.J., Minter, C.F. (2004). An ensemble-type Kalman filter for neutral thermospheric composition during geomagnetic storms. *Space Weather*, 2:1–9.

70. Reikard, G. (2011). Forecasting space weather: Can new econometric methods improve accuracy?. *Advances in Space Research*, 47:2073–2080.

71. Tsagouri, I., Belehaki, A., Bergeot, N., et al. (2013). Progress in space weather modeling in an operational environment. *Journal of Space Weather and Space Climate*, 3:A17.

72. Murray, S.A. , Bingham, S., Sharpe, M., Jackson, D.R. (2017). Flare forecasting at the met office space weather operations centre. *Space Weather*, 15:577–588.

73. Murray, S.A. (2018). The importance of ensemble techniques for operational space weather forecasting. *Space Weather*, 16:777–783.

74. Hapgood, M., Thomson, A. (2010). Space weather: Its impact on Earth and implications for business. *Lloyd's 360 Risk Insight*, London.

75. Austin, H.J., Savani, N.P. (2018). Skills for forecasting space weather. *Weather*, 73:362–366.

76. Mathew, J., Horbury, T.S., Wicks, R.T., et al. (2014). Ensemble downscaling in coupled solar wind-magnetosphere modeling for space weather forecasting. *Space Weather*, 12:395–405.

77. Schrijver, C., Kauristie, K., Aylward, A., et al. (2015). Understanding space weather to shield society: A global road map for 2015–2025 commissioned by COSPAR and ILWS. *Advances in Space Research*, 55:2745–2807.

78. Shprits, Y. Y., Zhelavskaya, I. S., Green, J. C., Pulkkinen, A. A., Horne, R. B., et al. (2018). Discussions on stakeholder requirements for space weather-related models. *Space Weather*, 16(4), 341–342. doi: 10.1002/2018SW001864

79. Food and Agriculture Organization (2018). *Impact of Early Warning Early Action*. Rome, Italy.

80. Guerra, J. A., Murray, S. A., Doornbos, E. (2020). The use of ensembles in space weather forecasting. *Space Weather*, 18(2), e2020SW002443. doi: 10.1029/2020SW002443.

81. Faulkner, R.W. (2012). Underground HVDC transmission via elpipes for grid security. *2012 IEEE Conference on Technologies for Homeland Security (HST)*, 359–364. doi: 10.1109/CITRES.2010.5619851

82. Qui, Q., Fleeman, J.A. (2012). Geomagnetic disturbance impacts and AEP GIC/harmonics monitoring system. In *Proceedings CIGRE USNC 2012 Grid of the Future Symposium*.

83. Kappenman, J. (2010). Low-frequency protection concepts for the electric power grid: geomagnetically induced current (GIC) and E3 HEMP mitigation. *FERC, Metatech Corporation*.

84. Zhu, H., Overbye, T.J. (2014). Blocking device placement for mitigating the effects of geomagnetically induced currents. *IEEE Transactions on Power Systems*, 30:2081–2089.

85. Tarditi, A.G., Duckworth, R., Li, F., et al. (2019). *High Voltage Modeling and Testing of Transformer, Line Interface Devices, and Bulk System Components Under Electromagnetic Pulse, Geomagnetic Disturbance, and other Abnormal Transients*. Oak Ridge National Lab.(ORNL), Oak Ridge, TN.

86. Bolduc, L., Granger, M., Pare, G., et al. (2005). Development of a DC current-blocking device for transformer neutrals. *IEEE Transactions on Power Delivery*, 20:163–168.

87. Etemadi, A., Rezaei-Zare, A. (2015). Optimal blocker placement for mitigating the effects of geomagnetic induced currents using branch and cut algorithm. *IEEE 2015 North American Power Symposium (NAPS)*, 1–6.

88. Bolduc, L., Granger, M., Pare, G., et al. (2014). Optimal placement of GIC blocking devices for geomagnetic disturbance mitigation. *IEEE Transactions on Power Systems*, 29:2753–2762.

89. Foster, J., Jihn, S., Gjelde, E., et al. (2008). *Report of the commission to assess the threat to the United States from electromagnetic pulse (EMP) attack. Critical National Infrastructures Report*.

90. Kovan, B., De Leon, F. (2015). Mitigation of geomagnetically induced currents by neutral switching. *IEEE Transactions on Power Delivery*, 30:1999–2006.

91. Pirjola, R. (2005). Averages of geomagnetically induced currents (GIC) in the Finnish 400kV electric power transmission system and the effect of neutral point reactors on GIC. *Journal of Atmospheric and Solar Terrestrial Physics*, 67:701–708.

92. Araejarvi, E., Pirjola, R., Viljanen, A. (2011). Effects of neutral point reactors and series capacitors on geomagnetically induced currents in a high-voltage electric power transmission system. *Space Weather*, 9: S11005.

93. Kazerooni, M., Overbye, T. J. (2019). Mitigating power system response to GICs in known networks. *Geomagnetically Induced Currents from the Sun to the Power Grid*, 219–232.

94. Orquin, A.R., Ramirez, V. (2016). Blocking geomagnetically induced currents (GIC) with surge arresters. *CIGRE US National Committee 2016 Grid of the Future Symposium*.

95. Hussein, A.A., Ali, M.H. (2016). Suppression of geomagnetic induced current using controlled ground resistance of transformer. *Electric Power Systems Research*, 140:9–19.

96. Ma, X.L., Weng, J., Liu, L.G., et al. (2010). Simulation study on converter transformer saturation characteristics due to GIC. *CICED 2010 Proceedings*, 1–6.

97. Johnson, M., Gorospe, G., Landry, J., Schuster, A. (2016). Review of mitigation technologies for terrestrial power grids against space weather effects.*International Journal of Electrical Power & Energy Systems*, 82:382–391.

98. Van Cutsem, T., Hassé, G., Moors, C., et al. (2004). A new training simulator for improved voltage control of the Hydro-Québec system. *IEEE PES Power Systems Conference and Exposition, 2004*, 366–371.

99. ABB (2012) *SolidGroundTM grid stability system. Geomagnetic Storm Induced Currents (GIC) and Electromagnetic Pulse (EMP) E3 protection*. ABB Inc. High Voltage Products.

100. Faxvog, FR., Fuchs, G., Wojtczak, W. et al (2017) HV Power Transformer Neutral Blocking Device (NBD) Operating Experience in Wisconsins. *MIPSYCON Conference November*, 1–15.

101. Kazerooni, M., Zhu, H., Overbye, T. (2017). Mitigation of geomagnetically induced currents using corrective line switching. *IEEE Transactions on Power Systems*, 33:2563–2571.

102. Klauber, C., Zhu, H., Overbye, T. (2017). Power network topology control for mitigating the effects of geomagnetically induced currents. *IEEE 2016 50th Asilomar Conference on Signals, Systems and Computers*, 313–317.

103. Lu, M., Nagarajan, H., Yamangil, E., et al. (2017). Optimal transmission line switching under geomagnetic disturbances. *IEEE Transactions on Power Systems*, 33:2539–2550.

104. Shao, W., Vittal, V. (2005). Corrective switching algorithm for relieving overloads and voltage violations. *IEEE Transactions on Power Systems*, 20:1877–1885.

105. Hedman, K.W., O'Neil, R., Fischer, E., et al. (2009). Optimal transmission switching with contingency analysis. *IEEE Transactions on Power Systems*, 24:1577–1586.

106. Lu, M., Eksioglu, S.-D., Mason, S.-J., Bent, R., Nagarajan, H. (2019). Distributionally robust optimization for a resilient transmission grid during geomagnetic disturbances. *arXiv preprint arXiv:1906.04139*.

107. Girgis, R., Vedante, K. (2013). Methodology for evaluating the impact of GIC and GIC capability of power transformer designs. *2013 IEEE Power & Energy Society General Meeting*, 1–5.

108. North American Electrical Reliability Corporation (2017). *Transformer thermal impact assessment White Paper. TPL-007-2, Transmission system planned performance for geomagnetic disturbance events*. North American Electrical Reliability Corporation (NERC), Atlanta, GA.

109. Foo, C.C., Chai, G.B., Seah, L.K. (2007). Mechanical properties of Nomex material and Nomex honeycomb structure. *Composite Structures*, 80:588–594.

110. Bachinger, F., Hackl, A., Hamberger, P., Leikermoser, A., Leber, G., Passath, H., Stoessl, M. (2013). Direct current in transformers: Effects and compensation. *e & i Elektrotechnik und Informationstechnik*, 1–5. doi:10.1007/s00502-012-0114-0.

111. Gurevich, V. (2011). Protection of power transformers against geomagnetic induced currents. *Serbian Journal of Electrical Engineering*, 8:333–339.

112. Elovaara, J. (2007). Finnish experiences with grid effects of GIC's. *Space Weather*, 311–326. Springer, Dordrecht.

113. Wamudson, M. (2016). *Operational Events in Off-Site Power System. Nuclear Power Plant Experiences and Mitigating Actions*. Energiforsk AB.

114. NERC Severe Impact Resilience Task Force (2012) *Severe impact resilience: Considerations and recommendations*. North American Electrical Reliability Corporation (NERC), Atlanta, GA.

115. Oughton, E., Copic, J., Skelton, A., et al. (2016). Helios solar storm scenario. *Cambridge Risk Framework Series, Centre for Risk Studies, University of Cambridge*.

116. NERC (2014) *TPL-007-1 transmission system planned performance for geomagnetic disturbance events*. North American Electric Reliability Corporation

117. Fleeman, J., Qui, Q., Anderson, L. (2013). Modeling, Monitoring and Mitigating Geomagnetically Induced Currents. *CIGRE US National Committee 2013 Grid of the Future Symposium*, 1–6.

118. Erinmez, I.A., Kappenman, J.G., Radasky, W.A. (2002). Management of the geomagnetically induced current risks on the National grid company's electric power transmission system. *Journal of Atmospheric and Solar-Terrestrial Physics*, 64:743–756.

119. Federal Energy Regulatory Commission (2013) *Reliability standards for geomagnetic disturbances*. Federal Register, Rules and Regulations

120. Semens, A.G. (2014). *GIC Module to Analyze Geomagnetic Disturbances on the Grid*. Siemens AG and Siemens Industry, Inc.

121. PowerWorld Corporation (2014). *GIC Modeling in PowerWorld Simulator*. Accessible on `http://https://www.powerworld.com/files/05DahmanGIC.pdf`.

122. Juvekar, G., Davis, K. (2019). MATGMD: A Tool for Enabling GMD Studies in MATLAB. *2019 IEEE Texas Power and Energy Conference (TPEC)*, 1–6.

123. Baril, G.A., Cahill, L., Dupont, A., Roberge, G. (1966). Commissioning of the first Manicouagan-Montreal 735-kV transmission lines. *CIGRE Report, 429*.

124. McGillis, D., Huynh, N.H., Scott, G. (1981). Reactive compensation. Role of static compensation meeting AC system control requirements with particular to the James Bay system. *IEEE Proceedings C (Generation, Transmission and Distribution)*, 128:389–393.

125. Théorêt, M. (1995). Système de mesure du décalage angulaire-Logiciel de l'unité centrale-Guide de l'usager. *Hydro-Québec internal report*.

126. Kamwa, I., Heniche, A., Cyr, C., et al. (2010). Power grid control research at Hydro-Québec: Recent advances enabling the development of technologies for a smarter transmission grid. *European Journal of Electrical Engineering*, 13:645–673.

127. Guillon, S., Toner, P., Gibson, L., Boteler, D. (2016). A colorful blackout: The Havoc caused by auroral electrojet generated magnetic field variations in 1989. *IEEE Power and Energy Magazine*, 14:59–71.

128. Trichtchenko, L., Guillon, S., Boteler, D., Pirjola, R. (2015). Responses of power systems in Canada to the space weather disturbances of the solar cycle 24. *EGUGA*, 7545.

129. Boteler, D.H. (2018). Dealing with Space Weather: The Canadian Experience. *Extreme Events in Geospace: Origins, Predictability, and Consequences*, 635–656. Elsevier, Amsterdam.

130. Bolduc, L. (2002). GIC observations and studies in the Hydro-Québec power system. *Journal of Atmospheric and Solar-Terrestrial Physics*, 64:1793–1802.

131. EIS Council (2014) *DC stray current reduction and blocking performance. Review of long term grid performance data*. EIS Council, New Zealand.

132. United States Government Accountability Office (2018). *GAO-19-98 Technology assessment. Critical infrastructure protection. Protecting the electric grid from geomagnetic disturbances*. United States.

133. Caruzzo, A., Belderrain, M.C.N., Fisch, G., et al. (2018). Modelling weather risk preferences with multi-criteria decision analysis for an aerospace vehicle launch. *Meteorological Applications*, 25:456–465.

134. Hapgood, M. (2018). Space Weather: What are Policymakers Seeking? *Extreme Events in Geospace: Origins, Predictability, and Consequences*, 657–682. Elsevier, Amsterdam.

135. Kox, T., Gerhold, L., Ulbrich, U. (2015). Perception and use of uncertainty in severe weather warnings by emergency services in Germany. *Atmospheric Research*, 158:292–301.

136. Federal Energy regulatory Commission, North American Electric Reliability Corporation (2019). *Joint staff white paper on notices of penalties pertaining to violations of critical infrastructure protection reliability standards.* Document no. AD19-18-000.

137. Jonas, S., McCarron, E. (2016). White house releases national space weather strategy and action plan. *Space Weather*, 14:54–55.

138. Jonas, S., McCarron, E., Murtagh, W. (2016). Space weather policy and effects. *IN-SIGHT*, 19:20–23.

139. Hapgood, M. (2015). Space weather: the public and policy. *Room Space Journal*, 1:106–110.

140. Pulkkinen, A., Bernabeu, E., Thomson, A., et al. (2017). Geomagnetically induced currents: Science, engineering, and application readiness. *Space Weather*, 15(7), 828–856. doi: 10.1002/2016SW001501.

141. Krausmann, E., Andersson, E., Gibbs, M., et al. (2016). *Space Weather & Critical Infrastructures: Findings and outlook.* EUR 28237 EN, Joint Research Centre (JRC), European Union.

142. EUR-Lex (2016). *Proposal for a Regulation of the European Parliament and the Council on risk preparedness in the electricity sector and repealing Directive 2005/89/EC (COM/2016/0862 final–2016/0377 COD).*

143. Karagiannis, G.M., Chondrogiannis, S., Krausmann, E., et al. (2017). *Power grid recovery after natural hazard impact. Science for Policy.* Report by the Joint Research Centre (JRC), European Union.

144. Perrow, C. (2007). Disasters ever more? Reducing US vulnerabilities. *Handbook of disaster research.* Springer,New York, NY.

145. Abi-Samra, N.C. (2010). Impacts of Extreme Weather Events on Transmission and Distribution Systems–Case Histories, Lessons Learned and Good Practices, Volume II. Electric Power Research Institute, Palo Alto, CA.

146. DGSCGC (2015). *Guide ORSEC Départament et Zonal: Mode d'action réatablissement et approvisionnement d'urgence des réseaux électricité, communication électroniques, eau, gaz hydrocarbures.* Direction Génerale de la Sècurité Civile et de la Gestion des Crises, Paris.

147. Abi-Samra, N.C. (2013). One year later: Superstorm Sandy underscores need for a resilient grid. *IEEE Spectrum*, 4:321–354.

148. Schmidt, M., Kelle, A., Ganguli-Mitra, A., De Vriend, H. (2009). *Synthetic biology: The technoscience and its societal consequences.* Springer Science & Business Media, Dordrecht.

149. Abi-Annan, K. (1999). *Report of the Secretary-General on the Work of the Organization General, Assembly Official Records Fifty-fourth Session (A/54/1).*

150. Hannigan, J. (2013). *Disasters without Borders: The International Politics of Natural Disasters.* John Wiley & Sons, Cambridge.

151. Katina, P.F., Keating, C.B., Gheorghe, A.V. (2016). Cyber-physical systems: Complex system governance as an integrating construct. In *Proceedings of the 2016 Industrial and Systems Engineering Research Conference.* Anaheim, CA: IISE.

152. Sheehan, N.T. (2010). A risk-based approach to strategy execution. *Journal of Business Strategy.* doi: 10.1108/02756661011076291.

153. Wilks, D. S. (2011). *Statistical Methods in the Atmospheric Sciences*, 100. Academic Press, Amsterdam.

# Part III

Developing a View of the
Geomagnetic Disturbances Risk

# 6 Geomagnetic Disturbance as a Perfect Storm

## CHAPTER CONTENTS

6.1 Natural hazard impact on power system.........................................................176
6.2 Geomagnetic disturbance as a specific natural hazard .................................180
6.3 Critical factors ...............................................................................................183
    6.3.1 Geomagnetic disturbance parameters ...............................................185
    6.3.2 Power system parameters...................................................................188
        6.3.2.1 Power system architecture...................................................188
        6.3.2.2 Power system operation state ..............................................191
        6.3.2.3 Power system grounding schemes........................................196
    6.3.3 Power system equipment parameters.................................................197
    6.3.4 Awareness ..........................................................................................198
6.4 High risk zones ..............................................................................................201
6.5 Economic loss estimation ..............................................................................205
6.6 Insurance perspective ....................................................................................213
6.7 Conclusion......................................................................................................218

ELECTRICITY for modern society is more than an amenity. It is essential for the national security, the support of national economy and the people's daily-life. Natural hazards are among the top threats to the electricity supply security. The power grid loss may result in great direct/indirect impacts on other infrastructures and have potential to severely influence the people's health. The physical processes associated with GMD appearance and its impact on modern technological systems are discussed in the first two parts of this manuscript. This part deals with the problem of the assessment of potential damages caused by the GMD from socio-economic perspective. It also focuses on the preliminary identification of high risk zones. It is believed that areas that have not been previously affected are particularly at risk. The assessment of the potential loss increases the awareness of the relevant stakeholders to the importance of the efforts dedicated to mitigate the risk. The detailed studies show to the society that "the zero risk level" does not exist. The hazard assessment is important for designing mitigation schemes. The evaluation of risk provides a basis for planning and allocation of limited resources: technical, financial and others.

## 6.1    NATURAL HAZARD IMPACT ON POWER SYSTEM

Both known hazards and a range of unforeseeable ones may pose a risk to power grid operation. Some of the threats are novel, while others have not been observed before. It limits the execution of the preparedness plans, since unforeseeable events can only be observed on the moment of recovery plan implementation. It increases the complexity and requires the search of new approaches for disaster risk assessment.

**Table 6.1**

**Overview of natural hazard effects in power systems**

| Hazard | Impact |
|---|---|
| High temperature & heat waves | power transfer limit decrease of the overhead transmission lines due to sagging |
| High winds during storms & hurricanes | fault or damage of infrastructure |
| Cold waves & ice rains | failures of the infrastructure. Ice accumulation on the isolators, which bridge the insulators provide a conducting path and result in a flashover fault. |
| Lightning strikes on or near overhead conductors | short-circuit faults, which trigger the relay protection and result in line disconnection. Normally, such faults are transient and the system rapidly recovers itself. However, the voltage surge caused by the strike may be transferred along the line and cause damage to power system equipment, e.g. power system transformers. |
| Heavy rains & floods | pose the danger to substation equipment, e.g. switchgear, control circuits, etc. |
| Sea level rise | threatens the coastal assets, impacts the hydro power generation and affects water availability for cooling purposes in thermal and nuclear power plants. |

Various definitions of disaster, hazard and risk exist. The ones given by the United Nations Office for Disaster Risk Reduction (UNISDR) are used in this monograph. *Disaster* is a serious disruption of the functioning of a community or a society at any scale due to hazardous events interacting with conditions of exposure, vulnerability and capacity, leading to one or more of the following: human, material, economic and environmental losses and impacts. *Hazard* is a process, phenomenon or human activity that may cause loss of life, injury or other health impacts, property damage, social and economic disruption or environmental degradation. In this terminology, space weather in the form of GMD is considered as a hazard and blackout caused by GMD is a disaster. *Risk* is an expected loss due to a particular hazard for a given area and a reference period. Following the terminology, natural hazard would be considered as a hazard caused by nature or natural processes.

Natural hazards have a potential to cause a physical damage to power grids. The failure scenario is predetermined by hazard type. The power grid response, in its turn, strongly depends on the physical vulnerabilities of its constituent assets, i.e. on their structural characteristics [1]. The extreme weather effects on the power grid include are described in Table 6.1.

The risks associated with different types of natural hazards are normally evaluated using different procedures leading to incomparability of results. Meantime, different natural hazards may occur at the same time or be triggered by each other. The potential consequences associated with specific hazard, while considering their interactions may result in a situation, when their combination is much greater than simply the sum of their parts [2]. Nepal M7.9 earthquake on April 24, 2015 is a typical example of a multi-hazard event. Multi-hazard event occurrence is assessed as a multi-risk process, within which the total risk from several hazards and interactions between them together with corresponding vulnerability levels are studied.

**Table 6.2**

**Major outages worldwide caused by natural hazards based on the number of affected customers**

| Date | Territory | Hazard type | Customers no. |
| --- | --- | --- | --- |
| July 2014 | Luzon, Philippine | Typhoon Rammasun | 13,000,000 |
| October 2012 | Eastern area, US | Hurricane Sandy | 8,100,000 |
| September 2018 | Hokkaido, Japan | Hokkaido Eastern Iburi earthquake | 2,950,000 |
| June 2012 | Ohio, US | Thunderstorm | 3,800,000 |
| September 2016 | South Australia | The Blath Tornado | 1,700,000 |
| November 2015 | Vancouver, Canada | Windstorm | 700,000 |
| February 2013 | Northeastern area, US | The North American Blizzard | 650,000 |
| December 2018 | Lower Mainland, Canada | Windstorm | 600,000 |

Natural hazard can be categorized as well as the triggered events associated with them (Fig. 6.1). Three weather conditions can be distinguished: normal, adverse, extreme, as a function of the consequences severity [3]. Faults due to the normal weather conditions have no significant effects. On the contrary, adverse and extreme conditions result in severe outages. Extreme event classified by Edison Electric Institute is an interruption resulting from a catastrophic event that exceeds design limits of the electric power system. According to [4], an event is classified catastrophic, if it leads to economic losses or casualties in excess of the following thresholds: USD 99 million total economic losses, 20 dead or missing, 40 injured and 2,000 homeless.

The list of severe blackouts based on the number of customers affected based on open-source data is given in Table 6.2.

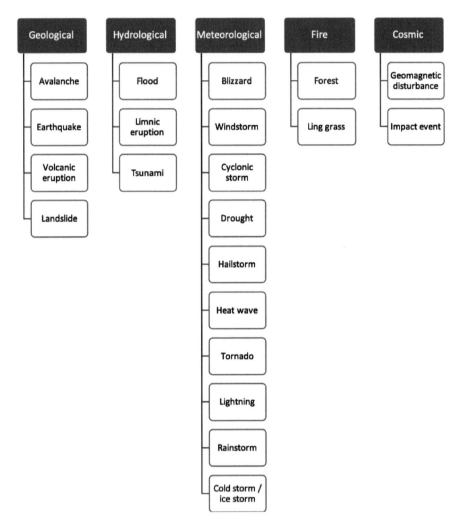

Figure 6.1: Natural hazard types

Extreme weather and climate events are among the primary causes of infrastructure damage causing large-scale cascading power outages, or shifts in the end-use electricity demands leading to supply inadequacy risks [5]. Severe weather events caused approximately 80% of the large-scale outages from 2003 to 2012 [6]. The annual loss in the US from such extreme events ranges from USD 20 to 55 billion [7]. European grid experiences similar pattern of disturbances. 30–60% of outages are weather related [8]. The restoration may take weeks, months or sometimes even years [9]. The graphs for outage duration are heavy-tailed and right-skewed, where the part in the extreme tails represents a catastrophic event impact.

Analysis of the number of events per year and associated losses in the past 40 years showed the rising trend (Fig. 6.2). The data is obtained from NatCatSER-VICE database provided by Munich Reinsurance Company (Munich Re), which is one of the three comprehensive natural hazard databases together with Sigma of Swiss Reinsurance Company (Swiss Re), and EmDat from Centre for Research on the Epidemiology of Disasters. The NatCatSERVICE employs around 200 sources that have been identified at a first rate for a particular region and/or type of event [10]. The main source groups consist of insurance industry information; meteorological and seismological surveys; reports and evaluations by aid organizations or NGOs, governments, the EU, the UN, the World Bank and other development banks; scientific analyses and studies; news agencies. Despite the fact of better reporting of much smaller events in the last years, the trend character is veritable. The analysis of Sigma database which is focused just on the catastrophic events gives the same trend.

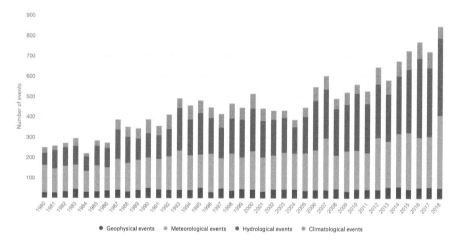

Figure 6.2: Number of relevant natural loss events worldwide 1980–2018 [Credit: NatCatSERVICE, Munich RE]

Terrestrial hazards were the scope of various studies. Following articles were devoted to extreme hydro-climatological events such as storms, wildfires, droughts, heatwaves [11], [12], [13], [14]. The barriers exist for improving the power grid resiliency to natural hazards: aged infrastructure, new requirements for electricity infrastructure, existing interdependencies between different power grids and a multitude of existing and emerging risks. The complexity of power grids challenges the search for most suitable method or combination of methods to improve its resilience.

World Economic Forum (WEF) develops a risk matrix on yearly basis in which top five risks are given in terms of likelihood and impact. Global risks are not strictly comparable across years, as definitions and the set of global risks have evolved with the new issues emerging on the 10-year horizon [15]. The ranking of risks "extreme weather events" and "major natural disasters" is constantly increasing over the years. GMDs are recognized as a new natural hazard of the modern technological age. It is important to mention that represented classification is a guide, not a definitive ranking of the risks. However, GMD impacts are relatively unfamiliar to the general public.

## 6.2 GEOMAGNETIC DISTURBANCE AS A SPECIFIC NATURAL HAZARD

Several decades ago, the GMD was just a scientific puzzle and curious phenomenon [16]. Meantime, major advancements in understanding the specific features of GMDs were achieved. However, there is a variety of questions, which are still open, especially, since we do have insufficient historical recordings to adequately examine the physical processes before, while and after extreme GMDs. The continuous infrastructure development of interconnected industries, which reliable operation is tightly dependent on electricity supply, makes GMDs to be considered as an emerging risk. Emerging risk can be classified as a peril that keeps risk managers up at night.

Contrary to more studied natural hazards such as earthquakes, windstorms and floods, power system planners and operators have little experience of power grid operation, while the GMD. It makes difficult to collect and systematize the data for revealing the correlations. GMDs have following specific features:

1. In contrast to terrestrial hazards, which have local and regional patterns and develop relatively slow, GMDs can be experienced on planetary scale within a few minutes.
2. GMDs simultaneously affect power grids over large distances (hundreds of square kilometers) by changing geomagnetic field parameters, which result in power system equipment and relay protection miss operation. Normally, mitigation guidelines are made for single point failures, which are more anticipated. Procedures for multi-point failures, particularly, when they are closely correlated in time, are not advanced.
3. GMDs impact results in power supply interruption on an area much lager than the hazard-stricken region itself.
4. GMDs rarely result in physical damage, though they may lead to significant economic loss, which leads to insurance coverage uncertainties.
5. Contrary to other terrestrial hazards, geographic position does not ultimately determine the probability of disruption. As it is further shown that the power grid robustness to GMD is determined by a multitude of critical factors of different nature.
6. From the power engineering point of view, the ideal description of a GMD event is the spatial-temporal characteristic of the horizontal magnetic field. It is important to note that it is not the instantaneous magnetic field footprint that dictates GIC but the fluctuations in the field. Fluctuations can be very complex and localized [17].

7. The GMD frequency appearance depends on solar cycle phase. Close to solar maxima several GMD with the severity Kp=5 and higher may arrive within the period of several days. Extreme GMDs Kp=9 may occur several times per cycle. Observations show that extreme GMDs may occur even close to the solar minima (Fig. 6.3). The blue curve shows the cyclic variation in the number of sunspots. Red bars show the cumulative number of sunspot-less days. The minimum of sunspot cycle 23 was the longest in the space age with the largest number of spotless days.

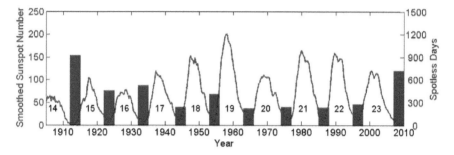

Figure 6.3: Sunspot cycles over the last century [Credit: Dibyendu Nandi et al.]

8. Among other geophysical hazards, an extreme GMD has the lowest occurrence probability rates compared to megaquakes and volcanic eruptions [18]. Results significantly differ depending on the considered set of historical events and used statistical models. Probability of a new Carrington-like storm in the next decade is between surprisingly likely and vanishingly unlikely.

9. GMDs cannot be visualized. Direct signal of GMD appearance on the Earth surface is received from sensors installed in geomagnetic observatories. Preventive signal is received from Advanced Composition Explore (ACE) satellite located in the first Lagrangian point (L1) about solar wind speed and direction. The ACE satellite is able to give a 15–30 minute warning as to whether a CME will hit Earth [19]. NOAA reports the likelihood and severity of the geomagnetic storms using the scale $(G1-5)$.

10. Similarly to other geophysical hazard, earthquakes, warning lead times for GMDs are very short and forecasting capabilities need to be improved. The time interval between GMD changes parameters of the Earth geomagnetic field and power grid blackout is too small for any meaningful mitigation actions. For example, the time interval was equal to 92 seconds for Hydro Quebec blackout.

11. The other difficulty is designated from the feature of GMD, which can be called as "arrival of series". The first moderate CME is normally followed by the second, which has higher speed and carries more energy. If the first CME is predicted 30–60 minutes in advance, the second CME,

which arrives on the heels of the first, is detected without any time gap for mitigation actions.

12. The GIC frequency is relatively small in comparison to the nominal power grid frequency (50/60 Hz). Referenced frequency range presented in the literature varies between $10^{-4}$ and $10^{-1}$ Hz.

13. The main difference between GIC and short-circuit current is the flow duration. Results of GIC registration in deadly grounded transformer neutrals showed that GIC amplitude is not constant and varies within the time period. Normally, moderate GICs of the amplitude of several Amperes flow in the period from few minutes to several days. High-amplitude GIC peaks appear for several minutes. The GIC fluctuation during the strong GMD Kp=7 registered in transformer neutral on substation 330 kV Vykhodnoy on June 1, 2013 is shown in Fig. 6.4. The GMD lasted from 3:00 to 6:00.

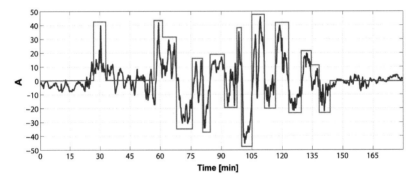

Figure 6.4: Geomagnetically induced current fluctuation in power transformer neutral on substation 330 kV Vykhodnoy during the strong GMD Kp=7 on June 1, 2013

14. The restoration time is determined by the failure scenario. When no equipment damage happened, restoration time is within 24 hours. Other cases are described in Table 4.2.

At present, a lack of recognition of GMDs as a driving force of grid disturbances exists. Several countries accepted the severe GMD impact on the national power grids. Nevertheless, moderate and severe space weather conditions are neglected that cannot result in power grid failure by themselves, but can be considered as a contributing factor among all others making power grid susceptible. The correlation between failures registered in NERC-DOE reports and solar activity was done in [20]. In contrast to the statement of NERC-DOE reports, which neglects GMD impact on power grid operation, ca. 50 grid disturbances were attributed to those, where solar activity played as a contributing factor. Similar to other natural hazards, severe GMDs may have significant economic and societal impacts (Section 6.5) including insurance industry (Section 6.6). The complete picture of pathways to failure is given in Fig. 6.5 (this figure is adapted from [21]), which is complimentary to Fig. 4.3, which describes an overview how GMD impact can evaluate to the blackout.

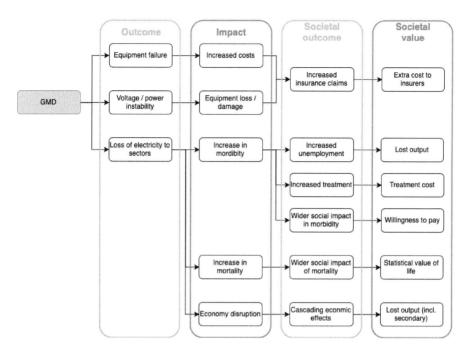

Figure 6.5: Geomagnetic disturbance socio-economic impact pathways

The perfect storm can be not just a severe GMD by itself. Severe societal-economic impacts can be derived from a multi-hazardous scenario. For instance, birds, lightning, earthquakes, over-drawing of power, failure of old infrastructure, overheating in heat waves, collapse of transmission lines in ice storms and instability caused by dead ends in the network could be other factors combined with GMD that lead to hazardous consequences [22]. This scenario was so far out of any research scope.

## 6.3  CRITICAL FACTORS

The findings presented in this book showed that the problem is multi-criterial. Historically, the problem of disastrous GMD impact on power grids was believed to be relevant only for high-latitude regions. In subsection 2.3.1 in Chapter2, the historic overview of GMD impact on power grid operation showed that the evidence was registered in a range of latitudes. The number of regions affected by negative GMD effects is growing, which means power grids located in mid-latitudes also experience state fluctuations even due to less strong GMDs. Additionally, GMDs may result in postponed outage, which are not always correctly correlated with the GMD activity. In other words, the trend of "high-risk" zones expansion exists.

There are disagreements among the studies about the level of risk posed by GMDs. The varying conclusions are a function of the list of critical factors that determine the amount of damage caused by GMDs. A general overview of critical factors is given in [23] that includes geomagnetic latitude, system topology (line length and orientation), line resistance and the transformers type. Nevertheless, the ranking of extended list of critical factors with respect to their impact was not given.

It is convenient to subdivide critical factors into four different groups (Fig. 6.6):

- **GMD parameters**: GMD type, ground conductivity, probability of appearance, geographic location.
- **Power grid parameters**: architecture, voltage level, power system state, geographic location.
- **Power system equipment parameters**: voltage level, equipment type, construction scheme, isolation characteristics.
- **Awareness**: social awareness, forecast, preparedness.

While much is known about these factors as a standing-alone player, there are gaps in understanding their interdependencies and their mutual role in GMD risk assessment.

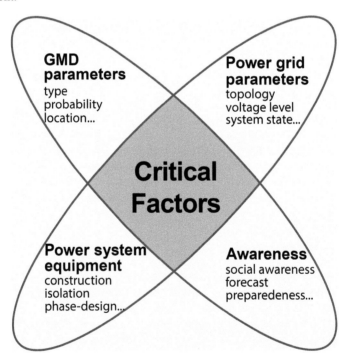

Figure 6.6: Group of factors that determine power grid robustness to geomagnetic disturbances

## 6.3.1 GEOMAGNETIC DISTURBANCE PARAMETERS

Each GMD is unique. Modern geophysics cannot precisely describe the morphological characteristics of different GMDs. Because the scientific data describing the hazard goes back only a few decades, and technology for space weather observation improved immensely over time. Any correlation of the data recorded in the different solar cycles is limited. Sometimes GMDs are described as "if you have seen one storm, you have seen one storm" [24].

Observations proved the concept that it is time derivative of the magnetic field that plays an important role in setting the GIC value, not the amplitude of the magnetic field variation. Nevertheless, magnetic derivative for each GMD is unique. The GIC amplitudes measured in the earthing lead of the 400 kV power transformer at the Rauma station, Finland, during two different storms are given as an example. GIC of only 40 A was measured during the Halloween storm, which is much lower than the value of 200 A measured in March 1991 [25]. The features of these two storms were quite different, though Halloween storm was stronger in terms of magnetic field amplitude.

Understanding two following physical processes gives the knowledge on drivers of hazardous GMDs: interplanetary solar-wind structures that drive hazardous GMDs and the response of magnetosphere-ionosphere system under hazardous conditions of solar wind. Exactly what is driving the development of complex magnetosphere-ionosphere processes and their connection to the development of extreme GICs is still open to debate [26]. One of the difficulties is that several mechanisms can be contributing at the same time [27]. Substorms associated with the enhancement of westward electrojets have been recognized for a long time as one of the most geoeffective causes of large-amplitude geoelectric fields in high latitudes [28], [29]. Lower attention is paid to eastward electrojets, though they can result in local field enhancement. Extreme local field enhancements may appear during severe storms, that is the local peak at least 100% larger than the regional distribution. The localized events were registered at different geomagnetic latitudes ranging from 50° to 85° N at time instances covering a wide range of local times; however, the physical processes that are responsible for the existence of these local extremes in the geoelectric field are unknown [30]. Intense geoelectric fields in equatorial zone are generated by the equatorial electrojet caused by rapid solar wind changes [31].

The dramatic GIC effects are associated with high $dB/dt$ values, which are more common for high-latitude regions. Some recent studies have shown that the equatorial boundary of the high GIC threat region lies between 50° and 60° magnetic latitude [32], [31], [33]. However, GIC impacts to power systems have been observed for small rates of 100 nT/min [34]. Sharp increases in $dB/dt$ are not limited to the onset of substorms, but can also occur with storm sudden commencements (SSCs) [35], [36], geomagnetic pulsations [37], [38] and sudden impulses (SIs) [39]. Although the $dB$ associated with SSCs and SIs are small compared to the $dB$ observed during substorms, the $dB/dt$ can still be large enough to inject GIC into the grid. SSCs are globally observed as a clear sudden change in the H-component magnetic field observed by ground-based magnetometers. At lower latitudes, the H-component enhancement resembles a stepwise increase caused by the sudden enhancement of the magnetopause current associated with the SSC, whereas at high latitudes the magnetic perturbation has a two-pulse structure [35]. The stepwise increase dominates at the equator and decreases moving poleward [40], and the two-pulse structures dominating at higher latitudes can sometimes be evident at low-to-middle latitudes as well [41], [42]. The GIC risk to power systems was quantified by specifying how often SSC events produced a $dB/dt$ of $>100$ nT/min [35]. Stations located at $<60°$ N MLAT observed an increase of $dB/dt$ larger than 100 nT/min in 2% of SSCs. The threshold exceeded for $>5\%$ of SSCs for stations located in the 64° N and 74° N MLAT range and for $>10\%$ of SSCs for stations located in the 64° N and 66°N MLAT range.

Understanding the extremes and their return periods are the important components of the GMD mitigation. Worst-case scenario can be defined in two ways: the return time of events of a particular intensity, or the largest expected event in a given time scale [43]. The analysis of extreme Dst index values was performed by [44], [45]; daily $Aa$ index [46]; half-daily $aa$ index [47]. However, the time rate-of-change of the field is widely regarded as the most relevant quantity for analysis of GMD impact on power grids [48]. The range of variations that may be observed up to a return period of 200 years in Europe was presented in [49]. Two conclusions were drawn: the predicted return magnitudes increase with latitude; the more northerly observatories can experience smaller extremes than those predicted around 55–60°. The range of extremes found in the analysis for 1-in-100 years and 1-in-200 years scenario is following. Horizontal field changes may exceed 1,000–4,000 nT/min in 1-in-100 and 1,000–6,000 nT/min in 1-in-200 year scenario at mid-European latitudes (55-60° geomagnetic). These numbers can only be considered as a guide value.

It is generally agreed within the space weather community that the Carrington event is the most hazardous GMD on the record. Evidently, there is no solar wind measurements for this event. The 3-D magnetohydrodynamics model of the Carrington event presented in [50] showed that the electric field of 26 V/km might be reached. It is comparable with the predicted theoretical maximum of 1-in-100 year scenario equal to 20 V/ km [32]. The distribution of the peak geoelectric field as a function of latitude is given in Fig. 6.7. The evaluated fields are more than two times larger than those generated during Hydro-Québec event (ca. 6 V/km) and Halloween storm (ca. 12V/km).

Extreme GMDs are characterized by the strongly shifted geomagnetic latitude boundaries. The minimum geomagnetic latitude, where aurora was observed, ranges between 18° N and 48° N according to simulations [51], [32]. The latitudes for other extreme events are comparable, e.g. 28° N for October 1870 storm, 20° N for February 1872 storm, 30-35° N for September 1909 and May 1921 storms together with Halloween event, 40° N for Hydro-Québec event [52], [53], [54], [32]. Overall, many world financial centers including London, New York, Frankfurt, Paris, and others are located within these latitudes. Though, the analysis of insurance claims filed by commercial organizations to Zurich Insurance over the 11-year period between January 2000 and December 2010 showed no significant dependence of the claim frequency associated with the geomagnetic latitude [55].

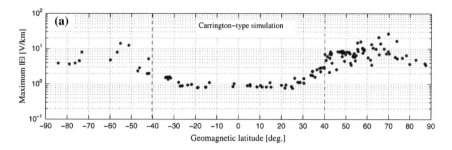

Figure 6.7: Global distribution of the peak geoelectric field determined for the Carrington-event type simulation [50]. The vertical red dashed lines refer to the locations of the transition regions between the middle and high latitudes [31]

It is important to note that the definition of hazardous parameters differs with the technology development. Even a less extreme event in terms of magnitude can lead to undesired consequences in modern critical infrastructure operation. According to Kp and Dst indices, GMDs can be described as in Table 4.1. The list of physical parameters that describe the storm's severity is given in [56]. These parameters are: massive soft and hard X-rays, CME traveling rapidly away from the Sun, the creation of large fluxes of accelerated particles, substational variations in a range of geomagnetic parameters, strong GIC. Severe space weather can be severe with respect to one parameter, but moderate or weak with respect to others.

Several studies addressed the probability of extreme event occurrence with different return periods. Limited time series make this task even more challenging. The opinions lay in the range "the 10-year recurrence probability for the Carrington type event is somewhere between vanishingly unlikely and surprisingly likely" [18]. The absolute limit for Dst is approximately 31,000 nT, which represents the complete cancelation of the Earth's magnetic field at the equator [57]. The smaller limiting value for Dst equal to 2,500 nT was set by [58] as a physical cutoff for Dst.

Lognormal distribution, extreme value theory and the power law distribution are statistical models used in the recent literature to describe the probability of GMD. The summary of the methods is given in [59]. The power law probabilistic analysis of Dst index variations on 55 years long record gives a 12% probability of Carrington size event (Dst < 850 nT.) every 79 years and estimate is reduced to 1.1% for more significant threshold of 1700 nT [60]. This number is considered somewhat too high, since the solar activity of that period was relatively high. The cumulative distribution function of magnetic storms using 89 year data set gave two times smaller probability of 4–6% of the Carrington type event for the 24th solar cycle [61]. The likelihood that a Carrington-sized event could take place 1.13 times per century is given by [62]. Analysis of the total accumulated exposure to E-field over the course of each event for 41 years period (1974–2015) registered in middle-latitude observatory Hermanous, South Africa, showed that the one in a century event has already been observed [63]. The data sets gave the magnitudes of $\Sigma E = 9.4 \times 10^4$ mV/km min and $\Sigma E = 1.09 \times 10^5$ mV/km min for 1-in-100 and 1-in-200 events, whilst $\Sigma E$ for the April 1994 event is $9.5 \times 10^4$ mV/km min. Once per-century excellence probabilities are often used to define space weather standards [64]. Moreover, these numbers should be kept in mind by grid owners and operators whilst choosing new power system equipment.

Looking beyond statistical hazard maps, ongoing algorithm development [65], [66], [67] could enable time-series scenario mapping of individual magnetic storms, convolving a time-dependent map of ground-level geomagnetic disturbance, derived from ground based magnetometer data [68], [69] with a map of Earth-surface impedance. Such a project could be further developed into a real-time geoelectric mapping service of use mitigating interference to power grid operations [70]. The current status of hazard map development and algorithms for "high-risk" zones identification is provided in Section 6.4.

## 6.3.2   POWER SYSTEM PARAMETERS

### 6.3.2.1   Power system architecture

The power grid architecture parameters that define robustness to GMDs are: voltage level, transmission line length and geographical location, where voltage level and transmission line length are normally interdependent parameters. To a certain degree, geographic location is also driven by the disposition of generation resources and load clusters.

The idea behind the development of HV and UHV networks was to transmit electric energy over long distances with a minimum loss. Increasing voltage levels cause the average circuit resistance to decrease, which in turn results in smaller power losses. The most developed power grids in terms of transmission line length and used voltage levels are North Americans, Chinese and Russian power grids. The US power grid was expanded in the following way. The operating levels of the high voltage network has increased from the 115 to 230 kV levels of the 1950s to networks that operate from 345 kV, 500 kV and 765 kV across the continent [71]. The modern high voltage gird consists of three operating voltages 345 kV, 500 kV and 765 kV. The most of the network is represented by 345 kV lines (64% of total transmission lines length). The highest operating voltage is 765 kV and is primarily located in the Illinois, Ohio, Indiana, West Virginia and upstate New York regions of the US Both the 345 kV and 500 kV portions of the network are more widely distributed across the US In total, 30% of power transformers are installed at voltage levels 500 kV and 765 kV.

The State Grid Corporation in China plans to develop UHV power grid in order to link remote resources in China to the load centers. The designed grid will consist of a combination of 800 kV HVDC and 1,000 kV AC transmission lines and will be used for large-scale super-long-distance power transmission, e.g. 1 GV transmission line Shanbei – Jundongman-Beijing 1,170 km, 1 GV transmission line Leshan – Shanghai 2,000 km. The total distance of transmission lines will be 38,800 km. The grid is designed to deliver 370 GW of power from coal, hydro and wind power plants with 282 GW of exports between regions. China is the first nation with the stated ambition of using UHV lines as the core network to interconnect its regional power grids into a strong national system. The highest voltage level both for AC and HVDC before grid's renovation was 500 kV.

The generation and load distribution is also uneven in UPS of Russia. The high voltage interconnected grid was developed for achieving more economic benefits in power transfer, optimizing power flows, sharing peak capacity between different time zones. The Russian interconnected grid consists of seven synchronously working regional TSOs. The presented voltage levels are 220–750 kV. The most of the power transformers are installed at the voltage level 220 kV and result in total of 77% of installed units. Single phase power transformers are installed at the operating voltages 500 kV and 750 kV and amount in 14% of installed units.

The Scandinavian power grid is a detailed case study for GMD effects. First of all, the main part of Scandinavian grid is located in the aurora and subaurora regions. It has a strong historical record of technical failures caused by GICs [72]. The Scandinavian grid consists of three synchronously working power grids in Finland, Sweden and Norway. GMD parameters that affect the grid are similar or almost similar. Nevertheless, the Swedish grid experiences the strongest negative impact from GICs. It shows that other factors also predetermine power grid robustness to GMD. The developed model is based on the actual ENTSO-E model [73]. It has a realistic number of nodes and can estimate the real situation with an acceptable precision. The Scandinavian grid was also analyzed in other studies [74], [75]. The specific feature of presented study is that only the 400 kV network is considered, since GIC prefers to flow over the lines with low resistivity. In addition, the "polar transit" is taken into account as well (the polar transit is represented by 150–220 kV lines, which connect 400 kV substation Varangerbotn, Norway, and hydro power plants Pirttikoski and Petäjäskoski in Finland). The model comprises a total number of 120 nodes. The equivalent calculation scheme is represented in Fig. 6.18 (see Appendix of this chapter). The required data for modeling is as follows:

- power grid data
  - network topology
  - nodes coordinates
  - transmission line parameters
  - characteristics of installed reactive power compensation equipment
  - installed power system equipment characteristics
- ground conductivity
- GMD parameters.

The 400 kV line resistivity is 0.008 Ω·m/km [76]. For the 220 kV and 150 kV transmission lines, the values 0.012 Ω·m/km and 0.016 Ω·m/km were taken in accordance with the physical dependences between the line's resistivity and voltage levels. The European Resistivity Ohm model (EUROHOM) is taken as a ground conductivity model. This model is based on extensive literature review and inversion of published magnetocelluric curves into a one-dimensional layered structure [77]. The exact value of permittivity is not significant due to the extremely small displacement currents compared to Ohmic currents [78]. GMD parameters depend on geomagnetic latitude. The location of grid's nodes as a function of geomagnetic latitude is represented in Fig. 6.19 (see Appendix of this chapter). The electric field conditions are modeled for hypothetical GMD with intensity $K = 9$ ($\Delta B = 500 - 1000$nT, $\Delta t = 3$ minutes) in the geomagnetic latitudes $48°$ N to $83°$ N.

The authors [79] have shown that reasonable understanding of GIC flow can be achieved even with incomplete data. The final model used here is described by the following set of characteristics:

1. The locations of nodes are given with varying accuracy up to 20 km. Nodes located within 20 km distance from each other are considered as a single node.
2. All power plants are included in the model.
3. All transmission lines are assumed to be straight if no detailed information was found. However, transmission line path is chosen in respect with real topology of the region. The real GICs are smaller than the calculated ones.
4. The node locations are given with varying accuracy up to 20 km. Nodes located within 20 km distance from each other are considered as a single node.
5. Transmission lines with installed series capacitors for reactive power flow regulation are represented as a circuit breakage.
6. All nodes are far enough from each other to prevent them from bearing each other [80].

The performed analysis shows that the network topology plays a more important role in GICs distribution than the geomagnetic latitude of the network if the geophysical parameters remain the same. High GIC levels are obtained in the nodes with big number of connections, which are longer than 100 km ($l_{TL} > 100$ km, where $l_{TL}$ is the length of the transmission line). In other words, the higher is the density of long high voltage transmission lines in the grid, the higher is its vulnerability. This dependence is more marked than the dependence on geomagnetic latitudes in the regions with constant ground conductivity, e.g. Scandinavia. Nonetheless, the transformer and grounding resistances also play a role in GIC amplitudes distribution. In case the line resistance dominates the transformer and grounding resistances, the GIC is defined by the line length (Eq. 6.1):

$$I_{GIC} \sim \frac{E}{\rho_L} \tag{6.1}$$

where $\rho_L$ is the transmission line's specific resistance (the line resistance in $\Omega$/km) and $E$ is in V/km.

Otherwise, even nodes at a shorter line length can experience high GICs. The highest GIC levels are obtained on the nodes in the southern Finnish grid, southern Norwegian grid and substations in the central Sweden. This result is correlated with [81]. The choice of calculation scheme, namely the choice of boundary points, significantly affects the accuracy of the results. Similarly, it was also shown by [82] that the ends and corners of a grid are prone to large GICs. In order to model the "end points" correctly, it is necessary to consider the GIC that can flow from/to neighboring grids rather than GIC flowing to ground through transformers at the edges of a network. It is possible to use an equivalent circuit for a neighboring network as shown by [83].

It tells that the ends and corners of the network are particularly prone to large GICs. Generators and generator step-up transformers tend to be located there. Therefore, the equipment installed in the nodes and mid- and low-latitudes may also experience destructive GIC impact. It is driven by the fact that the GICs are more related to the characteristic length of the whole network rather than to specific line lengths [84]. The impact of topology change on GIC distribution was also proved for the mid-latitude Uruguayan power grid [85]. The analysis results presented in this chapter show why the intensive development of Chinese high voltage power grid increases its vulnerability to GMD (Section 2.3.1).

The power grid sensitivity to GICs depends on the load situation, which is a constantly fluctuating value. It means that a certain GIC value that can be ignored in one system may be hazardous in another [86]. This dependence is studied in the next Section 6.3.2.2.

### 6.3.2.2 Power system operation state

The power system connection scheme is dictated by its operation state. The hypothesis that severe GMD impacts are more troublesome at light loads was stated by F. Koza in [87]. Nevertheless, a mathematical verification was not provided. The comparison of winter maximum and summer minimum power system states of the central Yakutia power grid is undertaken.

Any network can be represented by a graph with **N** nodes and **L** links. The calculation scheme for the Yakutia electric network is shown in Fig. 6.8 using graph theory. Note that the calculation scheme is consistent with the connection scheme. Dashed lines are switched off in the summer minimum power system state.

Power generation in the region is supported by the cascade of hydro power plant Vilijskaya ($P_{total} = 680$ MW) and thermal power plant Mirninskaya ($P_{total} = 160$ MW). Power system consists of two voltage levels: 110 kV and 220 kV. The power system nodes correlate with the graph nodes as described in Table 6.3.

In order to avoid total node failure, the nodes are designed to withstand contingencies with the minimum number of disconnected connections by implementing different bus schemes types, where bus scheme is an arrangement of overhead bus bar and associated switching equipment. The bus scheme type determines the network graph after the failure. Based on the voltage level, number of the connections and the importance of the connected load, various bus schemes are used. Both, double bus scheme and double sectioned bus scheme are typical for the voltage level 110–220 kV. The "double bus" stays for the scheme when each network element is connected to both of the buses via circuit breakers, which allows us to isolate the short-circuited element without loosing the full node operation. In case the number

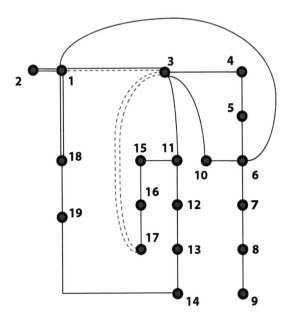

Figure 6.8: Network graph of Central Republic Yakutia power system [138]

of connections is high, each bus is sectioned for economical reasons using circuit breakers.

The analysis is done by taking into account three assumptions:

1. Due to size constraints three phase five limb power transformers are normally installed on the 220 kV substations.
2. It is assumed that the immediate loss of two power transformer units in one node occurs.
3. The actual limits for power flows over the transmission lines are not known. The maximum admissible values as a function of voltage level are considered.

Two power transformers are normally installed on the substations 35 kV and higher in UPS of Russia. The transformers installed power is chosen in accordance with $S = (0.6 \div 0.7)\, S_{max}$, where $S_{max}$ is the maximum power flow over network element. In this case, a power transformer is loaded on the level 60–70% in the normal state. In the emergency state, a transformer connected to the grid should ensure the customers power supply by taking into account the admissible thermal power transformer overload. Part of the lines is switched off in the summer minimum state due to decreased power demand. This allows us to consider disconnected power transformers as the back-up units, which are suitable for swift replacement. Thereby, such contingency can be considered as the normative one, that means the loss of $N-1$ element.

**Table 6.3**
**The characteristics of power system nodes**

| Node | Name | Voltage level, kV | Bus scheme |
|------|------|-------------------|------------|
| 1 | Substation Raionnaya | 220 | Double bus scheme |
| 2 | Vilijskaya hydro power plant | 220 | Double sectioned bus scheme |
| 3 | Sustation Gorodskaya | 220/110 | Double sectioned bus scheme |
| 4 | Substation NPS 13 | 220 | Sectioned bus scheme |
| 5 | Substation Olekminsk | 220 | Sectioned bus scheme |
| 6 | Substation Sunar | 220/110 | Double sectioned bus scheme |
| 7 | Substation Eligai | 110 | Sectioned bus scheme |
| 8 | Substation Kundyadya | 110 | Sectioned bus scheme |
| 9 | Substation Nurba | 110 | Sectioned bus scheme |
| 10 | Substation Taibohoi | 110 | Sectioned bus scheme |
| 11 | Substation Mur'ya | 110 | Sectioned bus scheme |
| 12 | Substation Sev. Nuya | 110 | Sectioned bus scheme |
| 13 | Substation Dorojnaya | 110 | Sectioned bus scheme |
| 14 | Substation Zarya | 110 | Sectioned bus scheme |
| 15 | Substation Lensk | 110 | Sectioned bus scheme |
| 16 | Substation Yarovslavskaya | 110 | Sectioned bus scheme |
| 17 | Substation Peledyi | 220/110 | Double sectioned bus scheme |
| 18 | Substation Mirnyi | 220/110 | Double sectioned bus scheme |
| 19 | MRGES | 110 | Sectioned bus scheme |

The power system scheme is going to look like the one in Fig. 6.9 in case of the loss of substation 220 kV Raionnaya and substation 220 kV Gorodskaya and like in Fig. 6.10 in case of the loss of power transformers in substation 220 kV Suntar. The region's power supply is going to be limited until the power system equipment recovery.

In order to perform the analysis, the network must be denoted in formalization firstly. The node in an undirected network is characterized by the degree $k_i$ of node $i$ (Eq. 6.2), which is the number of links $a_{ij}$ connected to node $i$, and clustering coefficient $C_i$ (Eq. 6.3) that shows to which nodes in a graph node $i$ intend to cluster together. Power grid parameters that correspond to the graph for winter maximum state are shown in Table 6.4.

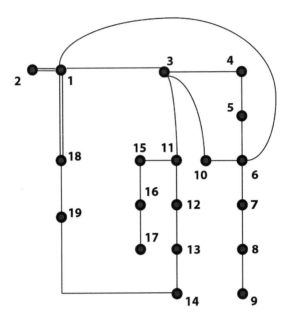

Figure 6.9: Network graph of Central Republic Yakutia power grid corresponding to summer minimum load normal state

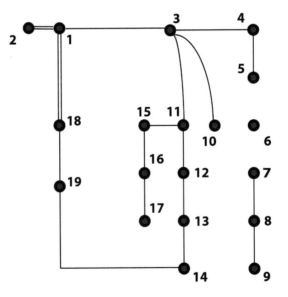

Figure 6.10: Network graph of Central Republic Yakutia power grid corresponding to summer minimum load state in case of substation 220 kV Suntar loss

$$k_i = \sum_j a_{ij} \qquad (6.2)$$

$$C = \frac{1}{N} \sum_N \frac{e_i}{k_i(k_i - 1)}, \qquad (6.3)$$

where $e_i$ is the number of links directly connecting neighbors.

**Table 6.4**

**The characteristics of graph nodes**

| Node | Nodes degree | Clustering coefficient | Node | Nodes degree | Clustering coefficient |
|------|------|------|------|------|------|
| 1 | 7 | 0 | 10 | 1 | 0 |
| 2 | 2 | 0 | 11 | 3 | 0 |
| 3 | 6 | 0 | 12 | 2 | 0 |
| 4 | 2 | 0 | 13 | 3 | 0 |
| 5 | 2 | 0 | 14 | 3 | 0 |
| 6 | 4 | 0 | 15 | 1 | 0 |
| 7 | 2 | 0 | 16 | 1 | 0 |
| 8 | 2 | 0 | 17 | 3 | 0 |
| 9 | 1 | 0 | 18 | 4 | 0 |
|  |  |  | 19 | 2 | 0 |

The topological characteristics analysis allows estimating the grid's vulnerability. There are lots of different definitions of vulnerability. In this book, the vulnerability is defined as the decrease of system's efficiency after the network element loss. The vulnerability of a network can be measured by the global efficiency of the graph describing the network. The global efficiency is a measure of the network performance. The network efficiency is based on the structure/ topology of the network. The aim of the global efficiency is simply to measure the average of the efficiencies over all pairs of the distinct vertices [88]. The grid's global efficiency $E$ is estimated using Eq. 6.4:

$$E = \frac{1}{N(N-1)} \sum_{i \neq j} \frac{1}{d_{ij}} \qquad (6.4)$$

where $d_{ij}$ is the distance, that is the length of the shortest path which connects nodes $i$ and $j$.

It is assumed that the immediate loss of two parallel power transformer units occurs in one node, e.g. the loss of power equipment in nodes 1, 3 and 6. Contingencies result in the grid scheme's change and, consequently, in grid's parameters change. The grid's efficiency as a function of a power grid state is represented in Table 6.5.

**Table 6.5**
**Grid's efficiency as a function of its operation mode**

| Scheme | Winter maximum load | Summer minimum load |
|---|---|---|
| Normal | 0.20245 | 0.19362 |
| Node 1 loss | 0.16496 | 0.18370 |
| Node 3 loss | 0.162016 | 0.15869 |
| Node 6 loss | 0.162037 | 0.149867 |

Node loss due to the contingency decreases the network's efficiency. Each node of the network is characterized by its unique contribution. The vulnerability $V_l$ of electrical network as a result of node $l$ loss is determined by Eq. 6.5:

$$V_e(l) = \frac{E - E_l}{E} \qquad (6.5)$$

The summer minimum load state is characterized by heavy transfer patterns via backbone long-distance transfers. Node loss may result in heavier voltage avalanche with minimal capacity for mitigation. At peak loads such as winter maximum, almost all generators are running. Consequently, a lot of spinning mass (i.e. total mass of the generators connected to the grid) for system stabilization is available, since the inertia in those spinning masses determines the immediate frequency response to inequalities in the overall power balance. For example, power equipment loss in node 6 (220 kV substation Suntar) degrades the grid's efficiency from 0.194 to 0.150 (rounded values), where the grid's vulnerability value is 0.225. Here, $E_i$ is the efficiency in case of the loss of node $i$. Calculated vulnerability from the same contingency at the maximum winter load is 0.199. In other words, the GMD impact at summer minimum load results in higher direct and indirect losses. The achieved dependencies are valid for the other power grids as well.

### 6.3.2.3   Power system grounding schemes

One more parameter that defines the grid vulnerability to GMD is the grounding scheme. All high-voltage power system elements are grounded because of safety requirements. Normally, the solid grounding scheme is used, where the power system equipment is directly grounded using a conductor of negligible resistance. Often the modification of grounding schemes tends to divert high GIC from one station to another. It is important to develop a possible geophysical scenario and power system operation states in advance, and to allocate the high GICs positions. Even the slight change in the grid's topology can lead to large variations in GIC flow and magnitudes [89]. The complete overview of the DC-blocking solutions installed in the ground wire and their impact on the overall power grid performance is given in Section 5.3.

Two different grounding schemes were tested, namely solid and resistive grounding. In the latter case, the conductor is a highly electrically resistive element for decreasing the GIC amplitude. For this purpose three cases **A**, **B** and **C** were calculated. Case **A** corresponds to the normal condition when all high-voltage neutrals are solidly grounded because of safety requirements. In Case **B**, resistors were installed in the top ten nodes with the highest calculated GICs. In Case **C**, resistors were installed in the top 10 nodes with the highest number of connections. Scandinavian power grid scheme presented in Subsection 6.3.2.1 was again the test scheme. On the grid level, the installation of resistors in transformer neutrals just redivides the actual GICs distribution, the GICs are not completely diminished. The undertaken analysis shows that the choice of nodes for installing resistors in transformer neutrals has to be done with respect to "P" typical power system states and "G" geomagnetic scenarios. Otherwise, resistive grounding can cause unwanted current growth. For instance, the GIC in power transformer neutral at 400 kV substation Pyhänselkä is equal to 0.139 p.u. in Case **B** (Case **A** is taken as a base) and 1.1 p.u. in the Case **C**. The numerical results are given using per-unit system (p.u.), which gives normalized values using a common base. At hydro power plant Aura, the GIC in power transformer neutral is equal to 0.25 p.u. in Case **B** (Case **A** is taken as a base) and 0.85 p.u. in the Case **C**. At hydro power plant Saurdal, the GIC in power transformer neutral is equal to 1.20 p.u. in Case **B** (Case **A** is taken as a base) and 0.87 p.u. in Case **C**. At hydro power plant Petäjäskoski, the GIC in power transformer neutral is equal to 1.33 p.u. in Case **B** (Case **A** is taken as a base) and 1.0 p.u. in Case **C**. The nodes selection can be done by using analytical dependencies between power system state parameters (y) and control parameters (x) using the fractional-linear functions [90]. The correlation is expressed in the form of the binomial function of Eq. 6.6:

$$x = \frac{aY + b}{cY + 1},\qquad(6.6)$$

where $x$ stands for complex current $\dot{I}$; $Y$ is the complex conductance of the power transformer grounding and $a$; $b$ and $c$ are complex constants. The analytical dependence of Eq. 6.6 can be adapted for any set of the typical power system states.

### 6.3.3  POWER SYSTEM EQUIPMENT PARAMETERS

Reliable power grid operation as a whole depends on the reliable power system equipment operation as a part. Power system equipment susceptibility to GMD is predominantly determined by its operation conditions, i.e. construction type. The relative comparison of power system equipment robustness to GMD effects is given in Table 4.8. The preliminary analysis of GMD impact is made by using the method of power quality weak link (Fig. 6.11). The principle runs as follows that the reliability of a grid depends on the susceptibility of the component, which has the smallest immunity mass. Even though the rest of elements may be capable of enduring severe power quality problems, a single element can render the entire grid extremely susceptible.

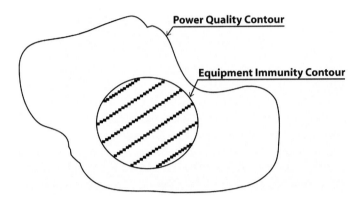

Figure 6.11: Criteria for power equipment quality

In case two elements work satisfactory by themselves, they might not function properly in cooperation. In such a situation, guidelines for minimizing the power quality interdependencies should be provided. The overview of mitigation action using operational procedures is given in Section 5.4.

### 6.3.4  AWARENESS

The authors in [91] states that people make judgements based on their general feeling about a situation. It leads to overestimation of frequently reported risks and underestimation of less frequently reported risks [92]. In general, four decision-making styles exist: vigilance, buck-passing, procrastination and hyper-vigilance [93]:

- Vigilant is a cool-hearted approach. It is viewed as a most effective style, which is a methodical approach utilizing a number of discrete stages which link clearly defined objectives to a consideration of a range of options with the final decision emerging from a careful assessment of the ramifications of each decision alternative [94].
- Buck-passing is a way of avoiding responsibility for a decision that has been made by suggesting that the decision is someone else's responsibility [95].
- Procrastination involves is the deliberate avoidance decision of taking in a given lapse of time [96].
- Hyper-vigilance model is often accused as "policy on the run". It is linked to substantial amounts of decision conflict or stress in the decision maker.

The Nevado del Ruiz eruption in Colombia in 1985 resulted in the tragic catastrophe of over 23,000 deaths, because local authorities and communities did not act on warnings [98]. The scientific knowledge was sufficient. It was shown that the impacts of the event can either be completely avoided or at least minimized by alerting the society to the threats and by raising awareness of preparation need including the possible evacuation [99]. Each disaster is unique, since it impacts the system with new properties as the result of evolutionary behavior [97]. Such behavior can be called emergent. Recently, disasters have shown a number of emergent effects. There is a strong need to develop new ways to prepare for unanticipated disasters.

Awareness as a critical factor consists of two aspects: social awareness of possible effects and mitigation actions, as well as forecast of GMD occurrence. In terms of resilience, increased social awareness helps to improve both technological (Section 5.1) and community resilience. Community resilience is a strategy to positive functioning and adaptation after a disturbance. In other words, community disaster resilience depends on the effectiveness of disaster management to ensure safety and well-being of the society. Community capacity to withstand the disaster is determined by the cultivation and use if transferable knowledge, skills, systems and resources that affect community- and individual-level changes consistent with goals and objectives [100]. Community resilience set of capacities differs from the technological one and includes [101]:

- **Economic development**: fairness of risk & vulnerability to hazards, level & diversity of economic resources, equality of resource distribution;
- **Information and communication**: narratives, responsible media, skills and infrastructure, trusted sources of information;
- **Community competence**: community action, critical reflection & problem solving skills, flexibility & creativity, collective efficacy empowerment, political partnerships;
- **Social capital**: received social support, perceived social support, social connectivity (informal ties), organizational linkages & cooperations, citizen participation leaderships & roles (formal ties), sense of community, attachment to place.

Power systems have a long lifetime, so one should think of heightened awareness of rare events. One of the greatest challenges in raising awareness of low-frequency risks, such as GMD, is overcoming personal and organizational experience that has not been exposed to a severe event [102]. While some studies such as [103] argue that considerable differences exist, often in dependence of whether or not affected people live in hazard-prone areas [104], other studies conclude that there is little empirical evidence for such a proposition [105]. Other study [106] states that even countries with a perceived low domestic space weather risk can benefit from a global approach to mitigate space weather risks. Following examples prove this statement:

- Eruption of Eyjafjallajökull, 2010, when volcanic ash clouds restricted air transportation over 70% of Europe. The level of preparedness is one of the escalating factors in the crisis and, despite the existence of well-known precursors throughout the world, volcanic ash clouds were not included in the risk registers of many countries [107];
- Tsunami in Japan, 2011, the assessment of which was limited by pre-existing mitigation procedures of such an event, since tsunami risk was not identified as required. The study scenario did not take into account

the maximum historically recorded wave in the region, which resulted in underestimating the risk [108];

■ Hurricane Sandy, 2012, which affected the power grid equipment and utilities. In consequence, electricity undersupply became a driver of another crisis, which lasted up to 2 weeks and required White House to take mitigation actions.

Together with legislative actions and risk transfer mechanisms, information is a third pillar in hazard mitigation. Interaction between these three components is essential for efficient disaster risk reduction and contributes to the concept of resilience as part of proactive adaptation [109]. In other words, disaster awareness identifies a multitude of strategies and programs that need to be developed by emergency managers. Overall, three types of issues exacerbate community vulnerability: (a) the problems of understanding and using the enormous amount of available information; (b) isolation during a hazard event; (c) general lack of adequate awareness and preparedness [110]. Simply providing more information about the risk will not necessarily have any effect and may, in some circumstances, exacerbate the issue by pushing people further into denial [111]. [112] showed how people's relationships with information sources influence their interpretation of the value of the information. The question of optimal information delivery before and after the event is studied in [113]. Four vital lessons for extreme events include:

■ The need for resources for developing hazard knowledge and establishing potential threats via risk assessments.

■ The need for communication between the inherent scientific uncertainties in hazard management that leads to probabilistic analysis, which plays an ever-increasing role in crisis communication.

■ The value of providing warnings, typically through networks commonly known as early-warning systems.

■ The intricate role of decision-making supported by various tools such as digital maps, automated messaging and alerting tools, as well as new policies and procedures to communicate data knowledge [114].

An increased awareness does not necessarily result in better disaster preparedness. One of the bright recent examples is the Fukushima disaster in 2011. The Japanese government, the Nuclear Safety Commission (NSC) and the company at the center of the nuclear disaster, the Tokyo Electric Power Company (TEPCO) were aware of seismic, tsunami and nuclear risks [115]. Despite this, NSC did not put in place adequate regulations, consequently, TEPCO was not obliged to adapt to them. As a result, TEPCO addressed the crisis inadequately and even contributed to its worsening [116].

The GMD exposure is a worldwide threat and a concern is shared by many nations. The globalized technologies and systems are particularly vulnerable to GMD. The exposure area is not constrained by national border and may affect many parts of the globe simultaneously. No nation can withstand such exposure alone. Improved awareness among business community helps to improve understanding of the impact on technological systems used by business; reduce costs by limiting interruptions; improve maintenance procedures. The notion of *risk tolerability* suggests that there is a level of risk that people are willing to tolerate or accept. Defining a premium that the society will pay to avoid the negative GMD impact is difficult. Various approaches have been proposed. For instance, [117] suggests using the risk profiles to show the relation between exceedance probability and damages for various events and to choose among various management alternatives.

The number of countries that carries out operative space weather activities in the world has been growing significantly in recent years [118]. For instance, several initiatives took place in the countries, which were traditionally considered to have low GMD risk. A detailed description about the beginning of space research in Latin America, a full set of instruments available for Space Weather studies and corresponding activities can be found in [119], [120]. Mexico set the establishment of Mexican Space Weather Service (SCiESMEX) [121]. Brazil founded the Brazilian Study and Monitoring of Space Weather (Embrace/INPE) Program [122].

## 6.4 HIGH RISK ZONES

Disaster risk zoning is essential for its mitigation and preparedness strategies development. Geoelectric hazard risk can differ significantly from one location to another. Several studies exist which considered the risks of GMD activity on the country level, including Finland [123], Sweden [124], Russia [125], Spain [126], United Kingdom [127, 128], France [129], South Africa [63], Italy [130] and other countries. Analysis showed that even in the equatorial zone equipment is considerably more prone to GICs that traditionally thought [131]. The report on GMD impact across the United States [71] and regional North America work estimates the geoelectric fields in the United States and South Canada [132]. In support of NERC directive on GMD impact, preliminary geoelectric benchmarks for intense magnetic storms were developed [133]. Extreme-event geoelectric amplitude is mapped by combining extreme-event estimates of geomagnetic activity with the EarthScope magnetotelluric impedance tensors [134]. An example of sinusoidal variation at 240 s realized as a waveform over a duration of 600 seconds is shown in Fig. 6.12. Later, the hazardous geoelectric field distribution was also done for Pacific Northwest [135]. Calculated amplitudes in the Northern Midwest are comparable to those that brought down Hydro-Québec in 1989.

The significant progress in understanding the GMD risk across the national borders was taken within the European Risk from Geomagnetically Induced Currents (EURISGIC) project. Its main goal was to derive the statistical occurrence of the geoelectric field and GIC in Europe [136]. It was shown that on average GICs are clearly larger in north Europe than in central and south Europe [137]. Nevertheless, calculated GIC amplitudes do not give an indicator to the sensitivity of a power grid to adverse GMD effects.

One should keep in mind that extreme GIC parameters do not mean high risk. As it was shown before, the set of critical factors determine power grid susceptibility to

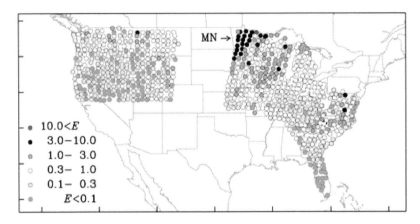

Figure 6.12: Geoelectric amplitudes (V/km) that can be expected to be exceeded once-per-century at EarthScope and USGS magnetotelluric survey sites for north-south geomagnetic variation with a period of 240 s and realized over a duration of 600 s [134]

GICs. One should not just look for the nodes where extreme GICs were observed or calculated, but to analyze the whole picture. In other words, one should look for the nodes with the high exposure.

The most widely used approach for identification of vulnerable power grids to GMD is described in [84]. The logic of it is as follows. The first step, called geophysical step, deals with the determination of geoelectric field associated with the GMD. The second step, called engineering step, is devoted to the GIC determination in the known conductor system due to the geoelectric field variation. The application of such an approach requires the knowledge of the geomagnetic field variation received from observatories closer to the grid, surface impedance and the detailed structure of the conductor system. These data are not available in the open source; therefore, the full set cannot be available for the calculation. Moreover, specific scientific knowledge puts further limits on its broad application. On the other hand, it is anticipated that the demand for detailed analysis of power grid's vulnerability to GMD will expand with increased awareness of the negative effects. Consequently, there is a clear need for a more accessible approach that can also be applied by stakeholders outside of the scientific domain.

The novel algorithm for preliminary evaluation of power grid's vulnerability to GMD is proposed in [138]. It consists of the steps illustrated in Fig. 6.13. In the first step (Step 1), data on the power grid has to be gathered and analyzed. This data covers information about installed equipment and their geographical locations, grid topology, voltage levels and lengths of transmission lines, and the ground's conductivity in the studied region. The second step (Step 2) requires the analysis of grid failure data, including the classification of causes. The failures caused by non-specified technical reasons have the potential to be the one triggered by the GMD. The correlation of

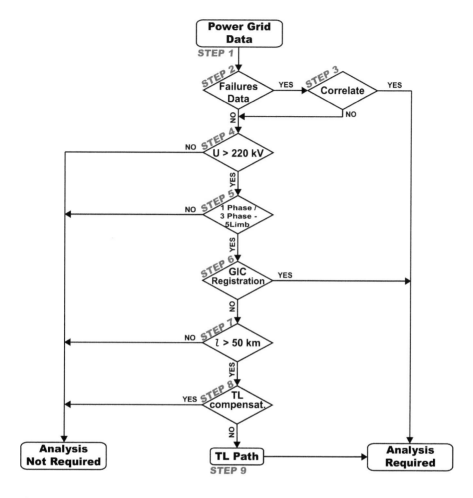

Figure 6.13: Algorithm to find power grid's bottlenecks to negative geomagnetic disturbance impact [138]

failures (caused by non-specified technical reason) with the moment of registered high $dB/dt$ is done within the third step (Step 3). If such a correlation was revealed, the detailed study of particular power grid robustness to GMD is required. If such a correlation was not revealed or could not be defined, the fourth step has to be taken. The fourth step (Step 4) involves the examination of voltage level of the installed power system equipment. If the power grid consists of nodes in which equipment with the voltage level higher than 220 kV is installed, the fifth step is necessary. If the voltage levels of all installed equipment is lower than 220 kV, then no detailed study is required. One can conclude that this particular power grid is robust to GMD effects. In the fifth step (Step 5), one has to inspect if any single phase or three phase five limb power transformers are installed in the grid. If the GIC recording equipment is installed in the vulnerable power transformer neutrals, the survey of the measured GIC amplitudes has to be done in the sixth step (Step 6). If the GIC levels exceed the threshold, the detailed study is required. The threshold is defined by the power transformer producer and has a unique value for each case. As it was shown before, the grid's topology plays an important role in determining the power grid's vulnerability. Therefore, the seventh step (Step 7) includes the investigation of potentially vulnerable equipment locations in the grid. If transmission lines longer than 50 km are connected to potentially vulnerable nodes, a further investigation has to be done. The further inspection (Step 8) of the grid's characteristics covers the search of transmission lines where reactive power compensation devices are installed. They help to maintain the voltage level within the admissible limits by minimizing reactive power deficit due to power transformer saturation caused by the GIC. The other effect is that series capacitors installed in the transmission lines, behave as circuit breakage from the GIC perspective. In the last step (Step 9), the data on ground conductivity in the region is summarized. The power grids located in regions with high resistivity crust have higher risks.

The advantage of this algorithm is that it takes into account a variety of critical factors of different nature, comprising power grid parameters, power system equipment parameters and awareness. Moreover, a least amount of data is required compared to the amount of data necessary for the classical approach. The reduced data set can be found in the open source. Whenever the grid is described as a potentially dangerous one, GICs in the network elements have to be determined. Thereafter, the stability of power grids particularly vulnerable to GMD effects can be studied using the classical methods of power system low-signal (small-scale) and dynamic (large-scale) stability analysis. It is important to mention that the proposed algorithm in this paper can be used by specialists who do not have a comprehensive understanding of the GMD effect on power grid operation.

The proposed algorithm is tested on the scheme of the Unified Power System (UPS) of Russia. UPS covers different geographical zones and is formed by seven interconnected power systems: East, Siberia, Ural, Middle Volga, South, Center and North-West, and they all operate synchronously. The territory covered by the UPS of Russia is spanning over nine time zones and includes different geographical regions located in high and middle geomagnetic latitudes. A considerable part of the grid is located in the severe climate conditions of the North and Far North. The total installed capacity is 235.4 GW (as of March 1, 2018). An important characteristic of the backbone grid is the large number of long transmission lines, namely 10,200 transmission lines of 110–750 kV. The total transformer capacity consists of 8,700 power substations of 110–750 kV among which 14% are single-phase units installed at 500–750 kV. Some regions are connected by weak intersystem links with reduced power transfer capacity.

The relative vulnerability of seven unified power grids of Russia to GMD effects is compared. The selected parameters for comparison reflect the choice of critical factors, that is, the installed power system equipment type and the equipment yielding, voltage level, transmission line length, grid topology, operation mode, geographical location, ground conductivity, transmission line path and data on power grid misoperation due to unidentified technical issues. These data are obtained and structured for all seven unified power systems [139]. The Siberian unified power grid is identified as the one which may experience the highest negative GMD impact. It mainly consists of weak interconnections of 220–500 kV. The "blocked" power of Sayano-Shushenskaya hydro power plant is equal to 2.4 GW. In addition, the 500 kV transit Gusinoozerskaya thermal power plant, Chita, is operated with decreased stability limit. The average 500 kV transmission line length is around 300 km. More than 50% of power transformers are maintained over the normative lifetime of 25 years.

The graphic representation of Siberian unified power grid is done in ArcGIS Map on the basis of National Geographic (2009). The following layers were introduced into ArcGIS: the physical map of the region, the power grid interconnection scheme, locations and parameters of generation and load nodes, and the ground conductivity map. The result of graphic visualization is shown in Fig. 6.14. The ground conductivity map is not reflected in Fig. 6.14. It was considered during the analysis of power grid robustness. Color graduation from red to orange corresponds to GMD risk minimization. Nodes planned for construction are marked in blue. Nodes, in which power transformer equipment was renovated in the past 5 years or more recent, are marked in violet. For instance, the main part of the grid of Yamalo-Nenets Autonomous Okrug was renovated in the last 5 years. This fact considerably increases its robustness to GMD effects though it is located in high-latitude region. The creation of such a model reduces the total cost for power grid (re)design and maintenance because it allows one to consider not only stability issues but also socio-economic and juridical limitations for power system development.

## 6.5   ECONOMIC LOSS ESTIMATION

There is still a great deal of uncertainty around the potential GMD impact on power systems and economic effect of such an impact. Despite it was shown to be essential for society well-being. High-risk zone identification associated with economic loss estimation is a stress test for industry, financial institutes, society and policy makers. Several stress tests were elaborated. An overview of them including GMDs scenario, modeling assumptions and main findings is given in this section. Overall, North America is the main studied region, with other areas being currently less in focus. The blackouts from other natural hazards can be used as proxies for the potential costs.

Business depends on reliable and high quality electricity supply for a variety of purposes. The electricity consumption grows faster than the population: 60% over 20% [140]. In the year 2018, electricity use grew nearly twice as fast as the overall demand for energy [141]. The vulnerability of electric power to an extreme geomagnetic storm remains the primary concern from an emergency management perspective [142]. Power system is a characteristic example of critical infrastructure. An infrastructure is termed critical if its incapacity or destruction has a significant impact on health, safety, security, economics and social well-being [143]. If electricity

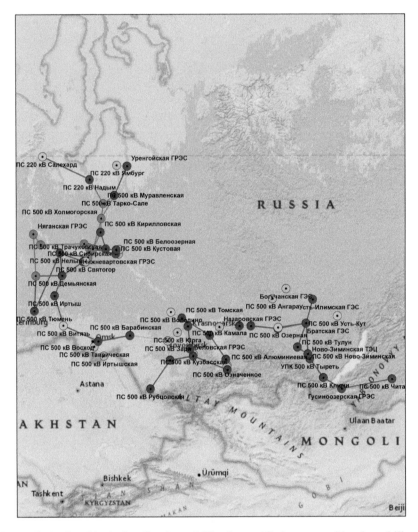

Figure 6.14: Graphical visualization of Siberian unified power grid vulnerability to severe geomagnetic disturbance. The color red stands for the nodes with a high risk, orange stands for nodes with a medium risk, blue stands for nodes planned for construction, and violet stands for nodes where the equipment was renovated not later than 5 years ago [138]

supply cannot, at least in short term, be substituted by another source of energy, it results in systematic failures across several infrastructures that have been created in interconnected way for increasing their efficiency.

Four critical infrastructure states are determined: operable, under threat, vulnerable and inoperable. Modern infrastructure networks do not operate in isolation, so they are interdependent, meaning the failure can propagate between the sectors. Several types of interdependency are distinguished [144]:

- **Physical**: two infrastructures are physically interdependent if the state of each is dependent on the material output(s) of the other.
- **Cyber**: an infrastructure has a cyber interdependency if its state depends on information transmitted through the information infrastructure.
- **Geographic**: infrastructures are geographically interdependent if a local environmental event can create state changes in all of them.
- **Logical**: two infrastructures are logically interdependent, if the state of each depends on the state of the other via a mechanism that is not a physical, cyber, or geographic connection.
- **Policy**: the change in form of a policy/procedure that takes effect in one part of the system influences another system.
- **Societal**: infrastructure operation is affected by public opinion.

Power grid failures caused by GMDs propagate and the impact interconnects with critical infrastructures operation via all above listed interdependency types. The study [144] also specifies failure types. They include cascading failures, when the failure in one infrastructure causes a disturbance in another infrastructure; escalating failures, when failure in one infrastructure worsens an interdependent disturbance in another infrastructure; and common cause failure, when two or more infrastructures are disrupted at the same time due to a common cause. It is difficult to anticipate the situations when simultaneous failures bring into play dormant and previously hidden interdependency pathways which destructively and synergistically amplify the failure [145].

The analysis of inter-infrastructural catastrophe is important for addressing (1) the processes and consequences of physical infrastructure failures in terms of physical capital losses and service flow disruptions; and (2) the resulting business disruptions and economic flow losses across the wider macroeconomic sectors [146]. Observations of inter-infrastructural catastrophe are rare and incomplete; therefore, the overview of elaborated catastrophe scenario is presented in Fig. 6.15. Some countries may have developed detailed models, but they are not in the open access. The report [147] presented in the German Parliament in April 2011 described a prolonged power outage in a large urban area. However, GMDs were not listed among potential causes.

The change of the state is determined by the backup power availability for each critical infrastructure specified in the legal norms which steers the color assignment in Fig. 6.15. The colors reflect different critical infrastructure operational states. For instance, the immediate localized degradation of Finance & Banking sector is driven by the dependance of financial activities on advanced technologies. Credit card processing, bank transactions, ATM, payroll disbursement, etc. are in the vulnerable group. Another example is the drinking tap water. As an example, Detroit had to abandon the drinking of mineral water for 72 hours after the restoration of power following the blackout in 2003. The threat of epidemic reached a critical level [148].

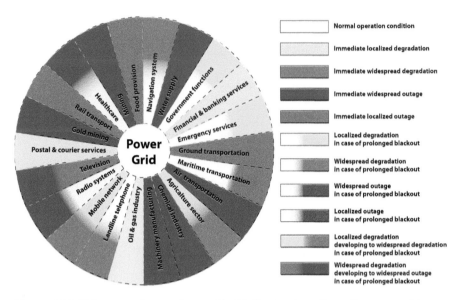

Figure 6.15: Primary and secondary critical infrastructure disruptions caused by severe geomagnetic disturbance

The overview of the whole water supply sector to GMD is given in [149]. The detailed overview of two Russian blackouts as a function of critical infrastructure is given in Table 6.6. Two major blackouts have happened in the recent Russian history: Moscow energy collapse on May 25, 2005, which was caused by a sudden thermal loss of 4 power transformers on 500 kV substation Chagino; St. Petersburg blackout on August 20, 2009 provoked by a differential relay protection mis-operation due to a communication cable sudden degradation on 300 kV substation Vostochnaya.

Primarily, GMDs affect electricity market. The evidence that the real-time market price in the PJM power grid in the US was affected by GMD was reported in [150]. It was also shown that the statistical relationship between a GIC proxy and various metrics of real-time operating conditions in PJM and 11 additional geographically dispersed power grids [151]. Later, it was commenced by noting the difference between the real-time and day-ahead price of electricity reflects unexpected operating conditions which are statistically related with a GIC proxy [152]. It was then proved that global indices do not present statistical correlation with electricity price due to aggregation and time difference, and electricity price and GMD parameters have non-linear correlation [153]. The severe events affect electricity market prices during the period of up to 2 days after the storm. Whenever the power grid fails, both electricity suppliers and customers are affected, though the character is different. Limited electricity supply results in the price growth, which results in the transfer of wealth. These transfers can be pretty large as the damage of the outages in California in 2001 was only USD 0.25 billion, but the transfer from electricity users to producers was ca. USD 40 billion [154].

**Table 6.6**

**Overview of blackout in Moscow on May 25, 2005 and St. Petersburg on August 20, 2009**

| Parameter | Moscow | St. Petersburg |
|---|---|---|
| Territory affected | 34 districts of Moscow region, Tyla region, part of Kaluga and Ryazyan regions | St. Petersburg districts located on the right bank of Neva river, 4 districts of Leningradskaya oblast |
| Disconnected power plants | 4 fossil-fuel power plants (2 GW) | 4 fossil-fuel power plants (2, 321 GW) |
| Oil & gas | 20 minute interruption of oil refinery | No data |
| Healthcare | Power loss in 28 hospitals | Power loss in 19 hospitals |
| Ground transportation | Traffic jam, subway degradation | |
| Government functions | Emergency power capacity used | |
| Emergency services | Emergency power capacity used | |
| Water supply | Interrupted for 25% of citizens | Interrupted for 40% of citizens |
| Railway transportation | Service interruption of 3 major railway stations for both regional and long-distance trains | Service interruption of 2 major railway stations for both regional and long-distance trains |
| Food provision | 2 milk manufactures interrupted their operation for 3 hours | No data |
| Mobile network | 6 million customers affected | 2 million customers affected |
| Internet | Number of unique users was decreased by 20% | No data |
| Financial & banking system | Banking system & Central Bank of Russia shifted the transaction time for several hours, Russian Trading system and Moscow Exchange stopped the operation | ATM degradation by 20% |

There is still no consensus on the correct approach for estimating economic loss. The method commonly known as input-output analysis has gained the most attention in recent years for its ability to model indirect or higher-order economic losses. Generally, the financial costs associated with power blackout can be divided into three categories: direct costs, indirect costs and resulting long-term costs of macroeconomic relevance [155]. The calculation of direct losses provides insight into the sale revenue change experienced by business in all economic sectors within the geographic territory affected by a GMD. The sector's indirect performance degradation is due to the reduction of the value added to the economy by the sector itself (profit, tax, etc.); the reduction in the sector's immediate consumption and demand for services and goods by other sectors and the reduction of sectors liability to meet demands of existing customers. In other words, both upstream and downstream indirect shocks are considered to propagate along the tier after the tier of suppliers and customers, respectively, which collectively constitute the supply chain of the economy [156]. The indirect losses value can vary within the interval limited by the losses caused by the production loss indirectly caused by other direct shocks (minimum bound) and loss in certain facilities who do not experience power shortage by themselves (upper bound). There is a split view on the way to calculate indirect or high-order economic losses. The most common is input-output approach, which spawned an entire field of related models, which include: the inoperability input-output model (IIM); Ghosh supply-side model; dynamic input-output models; key-linkages analysis; as well as inventory-based models amongst others [157]. World Input-Output Database (WIOD) is a unique data source which provides underlying data, covering 43 countries and 56 economic sectors [158]. It has high transparency over underlying data sources and methodologies used.

The cost of supply interruption is measured in two ways:

1. **Value of lost load (voll)** which is identified by [159] as the most reliable and robust proxy for estimating economic losses in relation to electricity supply security. The value can be minimized by cutting the customers with the lowest voll. Analysis done for the Dutch network showed that the economic cost is lowest when construction, the government and the the households are shut off as late as possible, and manufacturing would be shut down first [160]. It arises from the fact that the manufacturing uses 38% of total electricity but creates only 8% added value. Though technical operational constraints may limit this strategy. The voll differs also across time of the day and the day of the week.
2. **Damage per hour** which is used for deciding investments in network reliability. The investments are profitable in the regions where the value of damage per hour is high.

Two well-documented examples of economic loss caused by GMD are: the 1989 storm which led to voltage collapse of the Hydro-Québec grid (the total cost of damaged equipment was USD 6.5 million) and generator step-up power transformer loss in New Jersey. The net cost of Hydro-Québec grid failure is estimated to be USD 13.2 million [161]. The timeline of the events evolution is given in Section 2.3. Nevertheless, the evidence of blackouts caused by other hazards gives a basis for severe

GMD impact cost estimation. For instance, the Southwest blackout in 2011 left 2.7 million people without power for 12 hours or the June 2012 Mid-Atlantic and Midwest derecho impacted 4.2 million people across 11 states [162].

There is still an ongoing debate about the value of economic loss. Severe space weather events could lead to global economic damage of the same order as wars, extreme financial crisis and estimated future climate change. Overall, the economic impact of GMDs is the function of: (1) spatial and temporal extend of the hazard; (2) the vulnerability of the technologies/infrastructures susceptible to failure; (3) the degree of mitigation; (4) input production and consumption options available to firms and consumers [163]. It should be noted that any study on GMDs involves making assumptions or engineering estimates that limit the ability to accurately calculate the economic loss associated with the GMD risk to power grids. Though the critical factors are generally understood, the underlying assumptions, treatment of uncertainty, or depth or analysis differ. Nevertheless, studies share fundamentally similar methodologies for analysis.

Two reports give similar numbers in terms of the affected power transformers number, recovery period and economy impact. The century-scale GMD would cause a 10% of the power grid capacity reduction for up to a year. The impact could be between USD 2.4 and 3.4 trillion or 5.6% GDP. The effects would influence sectors and populations well outside the direct area of impact [87, 164]. Though [87] mentions longer recovery time up to 4–10 years. It is emphasized that the recovery time duration is more dependent on the damaged power transformer concentration than on the total number of damaged transformers [165].

The most radical view was given in [71]. The extreme GMD could create a loss of over 70% of the nation's electrical service. It was simulated that ca. 216 power transformers may receive GIC cumulative exposures. It is similar to the FEMA view that assumes a worst-case scenario until such time as the research and engineering matures sufficiently to provide high-confidence-level, high-fidelity impact forecasts. A worst-case scenario is currently defined as multiple states and regions without power for periods exceeding 2 days, with the potential for some areas to remain without power for weeks or even months [166].

The Helios solar storm suggested three catastrophe scenarios, named S1, S2 and X1, which differ from each other in the number of damaged transformers [156]. Direct losses affect 90 million US citizens (28% of the population). The largest losses are seen in services and manufacturing sectors, while government and agriculture sectors have smaller losses. The power will be fully restored within 6–12 months period depending on the scenario. Mining sector also experiences the largest indirect losses due to its importance in supplying products to other industries. Two highest indirect shocks for other countries for both upstream and downstream components correspond to China and Canada. However, Russia and India have higher than average between downstream and upstream components. Downstream component shows the US economy dependence on export and upstream component on import. Macroeconomic losses were studied using Oxford Economics Global Economic Model, which provides multivariate forecasts and describes the systematic interactions for the largest 47 economies of the world, with headline information on further 34 economies. Global GDP risk ranges from USD 140 to 613 billion across the three scenario variants. Depending on the scenario, the five-year baseline GDP projection on global GDP is within 0.03% to 0.26%.

Stakeholders should not be just aware of a catastrophic event. Vulnerability may arise due to continuing degradation as a consequence of many smaller impacts [167]. Isolated substorms causing a limited outage can also have significant economic impact [163]. Especially if no investment for national critical infrastructure hardening will be attributed, since aged infrastructure is more vulnerable to GMD effects.

Less critical views were given by JASON, an independent defense scientific advisory group, and Sandia National Laboratories. The main concern lies in the field of the plausibility of the worst-case scenario accepted in the previous studies [168]. The NERC GMD Task Force concluded that the most likely worst-case system impacts from a severe GMD event would be voltage instability and potential blackout. Furthermore, blackouts that originate in the transmission grid in the absence of substantial equipment damage are generally restored within 3 days and often much sooner, according to Los Alamos National Laboratory and NERC [169].

Though no in-depth analysis about the GMD impact on European economy was performed, the evaluation of other hazards socio-economic impact can represent the values. Large-scale blackouts still result in huge economic loss even if the power restored within few hours. The 2003 blackout is one most severe in the recent history that affected 56 million citizens over the Italian Peninsula territory and a part of Switzerland. The economic losses for Italy due to the 2003 power blackout as a critical infrastructure type function is given in [170]. Due to heat waves in Europe, nuclear power plants had to decrease their output in 2003. The reduced power value was 4 GW, which led to USD 14.5 billion financial losses [171]. The total direct loss with the highest damage factor is estimated exceeding Euro 10.3 billion (in 2006 prices) [172]. However, there is a lack of studies on long-term run effects.

Several factors define the economic loss from natural catastrophes. Countries with higher per capita income experience a similar amount of catastrophic events but suffer less death from these events [173]. Risk averse individuals will make different risk-return trade-off choices at different income level [174]. The latter reference identifies a number of social critical factors: nation with higher levels of educational attainment and greater openness for trade are less vulnerable to disasters; stronger financial sector and a smaller size of government (measured as the fraction of government expenditure per GDP) are associated with a lower disaster death toll. Institutional quality and international openness mitigate negative consequences [175]. Moreover, the poorer countries are also unlikely to be able to adopt the counter-hazard policies [176].

In addition, business will incur indirect costs through the need to establish operational procedures to monitor solar activity and initiate adaptation measures when needed [177]. Two types of costs are distinguished: *ex ante* – mitigating the catastrophe, and *ex post* – coping with the consequences [178]. Nevertheless, the economic benefits of mitigation could exceed the *ex post* cost [179]. Therefore, mitigation actions and steps underlined in Chapter 5 are doable and necessary.

## 6.6   INSURANCE PERSPECTIVE

Risk transfer from policyholders to insurance and later to reinsurance companies is an effective mechanism for financial stability ensuring. In other words, pooling risks reduces the uncertainty of expected loss over a given period of time. It helps to preserve the continuity of business, especially in the presence of extensive events. Babylonian King Hammarubi's code (1750 B.C.) was the first record of insurance. Across the centuries, merchants were forming a pool to spread the loss. Enhancing maritime information exchange in the Lloyd's Coffee House led to underwriting development. The first record is dated by 1757. Large fires across Europe drove the development of reinsurance sector: Cologne Re after the Hamburg Fire of 1842; and Swiss Re after the Glarus Fire of 1861. These severe events demonstrated the need for reinsurance, although many reinsurance companies were in fact founded to prevent the outflow of reinsurance premiums from local economies to foreign ones [180].

Risk of any event is subjective, and therefore includes a notion of uncertainty. Risk can be defined as the set of a given scenario, the probability of that scenario, and the consequences or the evolution measure of that scenario, where a scenario can be defined as an identifiable outcome [181]. This scenario is the basis of the catastrophe models, which are used for specifying the amount of risk to be transferred and the price for it.

Catastrophe model is a computerized system that generates a robust set of simulated events and estimates the magnitude, intensity and the location of the event to evaluate the amount of damage and calculate the insured loss as a result of a catastrophic event [182]. The identification of key catastrophe models development milestones is given in [183]. Catastrophe models are used by all three parties: insurance companies, who underwrites the original policy; reinsurance brokers, who works on behalf of the insurer to transfer risk to one or more reinsurances via reinsurance policies, and reinsurance companies. Depending on the needs of parties, catastrophe models have the functions described in [184], which are summarized in Table 6.7.

The graphical representation of a catastrophe model structure is given in Fig 6.16. Any catastrophe model consists of four components.

- *Hazard* is a process, phenomenon or human activity that may cause loss of life, injury or other health impacts, property damage, social and economic disruption or environmental degradation. Two types of hazard models are used: deterministic (scenario models), which are determined by assumed initial modeling conditions like historical hazard events or "what-if" scenario, or probabilistic models, which estimate the probability of event of a given severity. Contrary to deterministic models, probabilistic models can correlate spatial and temporal risks for a credibly scaled event. They combine historical data with theoretical and statistical models. The spatial distribution of critical hazard intensities is called footprint. Figure 6.12 is an

**Table 6.7**

**Catastrophe model functions**

| Insurance | Reinsurance broker | Reinsurance |
|---|---|---|
| Risk selection (i.e., **portfolio**) | Analysis of the client's insurer's portfolio | Design and price reinsurance structures |
| Price setting (i.e., **premium**) charged to the policyholder | Design reinsurance or retrocession structures | Allocate and monitor the maximum amount of business that can be written for individual business units |
| Allocate and monitor the maximum amount of business that can be written for individual business units (i.e., **capacity**) | Price reinsurance or retrocession structures | Target the most profitable combination of different contracts, given the company's constraints |
| Ensure that the business has enough capital to meet **solvency regulations** | Peer review client exposure and loss profile versus other insurers | Ensure that there is enough money held to pay back claims and keep the company solvent |
| Monitor adherence to stated risk guidelines and **risk policy** for key peril regions and business units | Provide insight into strength and limitations of different models | |
| Allocate reinsurance costs to different business segments, e.g. underwriting business units, branches, or even individual risks | | |

example of a footprint achieved from a deterministic model. In probabilistic modeling, footprint would also have a probability of footprint distribution. Normally, the main driver of a damage is chosen as a hazard metric based on which a hazard footprint is generated. Point observations, which are measurements of hazard parameters at a certain geographic location, are used as input data. The use of historical observations for hazard footprint modeling is not as straightforward as one could imagine. Three GMD scenarios are normally in focus: 1-in-10 years, 1-in-30 years and 1-in-100 years, though

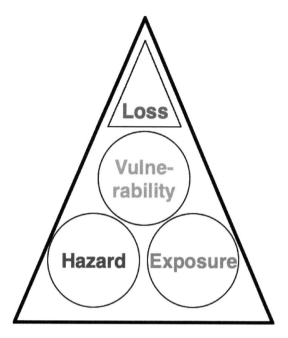

Figure 6.16: Catastrophe model structure

limited data points exist compared to other natural hazards. However, even for more frequent natural hazards, insurance companies have only 10 or 20-years time-series of claims. It is due to the fact of the underlying trends change such as exposure landscape, infrastructure reliability standards, construction/maintenance costs and others.

- *Exposure* data is the primary input to catastrophe model. The exposure can be evaluated using a combination of geospatial mapping and probabilistic modeling. The quality of exposure data varies greatly among parts of the world and the industry lines. For instance, data quality high resolution and rich attribute data exists for power plants, though the transmission grid is poorly represented. Together with object location and sum insured, primary and secondary modifiers are considered. The change of the exposure amount through the life-cycle is done by S-curve approach. Catastrophe models can assess the economic impact by associating an event-based model with an economic exposure database [185]. Three approaches are used for exposure database creation: bottom-up, top-down and Insurance-specific approach. Bottom-up approach evaluates location-level characteristics for every insurable property and aggregates by geographical region. Its main disadvantage is the limited data availability (inconsistent data vintages) and contingency issues. In addition, this approach is rather time consuming. However, it gives results with high resolution and can be described as user-independent. Top-down approach evaluates country-wide exposure and distributes to geographical regions with distribution factors. In comparence to bottom-up approach, it is quick and easy scalable. Consequently, it is coarse and inaccurate. An insurance-specific approach collects data from a proportion market participants by assuming that they are representative, scales values up to 100% using market share. It gives accurate market view, though following issues can be considered as its disadvantage: data mapping and data coherence, market share by premium different to sum insured, differences between companies in rebuild cost models.

■ *Vulnerability* model development involves quantifying the relationship between the severity of the hazard at a given location and the resulting loss [186]. In other words, vulnerability component links hazard and exposure components. Most vulnerability models are arranged as a series of damage functions, which enable look-up between hazard intensity and estimated damage as a ratio of total value [184]. Damage ratio $DR(i, j)$ for an $i$ objet from a $j$ peril can be represented as in Eq. 6.7:

$$DR(i, j) = \frac{\text{Repair cost } (i, j)}{\text{Total replacement cost } (i, j)} \tag{6.7}$$

The examples of damage functions for severe GMDs can be found in [71]. Two main approaches are used for vulnerability functions development (Table 6.8). In both cases, an expert knowledge-based input can be required.

---

**Table 6.8**

**Vulnerability functions development approaches**

| Description | Advantages | Disadvantages |
|---|---|---|
| **Empirical vulnerability approach** | | |
| Uses post-event studies as sources for damage information, often through regression analysis | - Derives the vulnerability function from data that are captured by insurance companies from their policyholders | - Geographic inaccuracy<br>- Need to define a typology or inventory classification<br>- Uncertainties associated with interpolation<br>- Highly dependent on data availability and quality |
| **Analytical vulnerability approach** | | |
| Use of computational and probability models mixture to measure the expected response to a peril | - Less-biased | - Sensitive to granularity of initial data |

- *Loss* module is an output of a vulnerability module. It translates infrastructure damage into costs covered by insurance. It is usually expressed as ground up loss which is the entire amount of insurance loss, including deductibles, before application of any retention or reinsurance; retained (client) loss, which is the loss to be insured; gross loss, which is the amount of a ceding company's loss irrespective of any reinsurance recoveries due. It is essential that the loss reflects the impact of insurance products and mechanisms. Loss module reflects any insurance and reinsurance policy conditions so that the impact of high-level policy structures are reflected and detailed at lower levels in the form of metrics: average annual loss (AAL), occurrence exceedance probability and aggregate exceedance probability (AEP) curves. Afterwards, the catastrophe model should pass the iteration validation process. At the end, there should be a clear understanding of:
  - Which spatial and temporal was achieved? Is this high enough?
  - How are footprints calibrated?
  - Which historical events in which territories are considered? For model creation? For its validation?
  - Which hazard metrics are used?
  - How does loss data vary among territories and industries?
  - What is the exposure data quality?
  - What are the sources and the range of uncertainties?

Currently, GMDs are not part of the classical business portfolio of the insurance industry. Several legislative steps were taken. European Directive on insurance regulation called "Solvency II" became effective on January 1, 2016. It states that any European insurance company should prove its financial sustainability to any knowable loss event with a probability of occurrence of 1-in-200 years. Earlier in 2015, the Bank of England Prudential Regulation Authority included a solar flare/GMD as one of the 11 stress tests recommended for insurers to estimate losses against it. It is not presented in the time of writing (spring 2020) among recommended stress tests [187]. Nonetheless, findings from other natural hazards may be used for the GMD impact evaluation. Lloyd's of London publishes on the annual basis realistic disaster scenario report in which space weather is considered. Though only space weather impact on satellites is studied.

Major insuring companies published reports on space weather impact on their business with the different level of details Lloyd's of London [165], Swiss Reinsurance Company Ltd (Swiss Re) [188], Aon plc [189] and Zurich Insurance Group Ltd. [190]. Nonetheless, no number of actual insurance business disruption is given in these reports. It is anticipated that awareness raised by the these publications may result in risk-mitigation actions implemented by the insured parties.

Contrary to other natural hazards, GMDs cause extensive economic damage by no or minor physical damage. [156] identified six categories of claimants such as power transmission operators, power generation companies, companies that loose electricity supply, homeowners, specialty (cancellation of public and private events). The described pathways in Section 6.2 show that one of the causes is the power transformer unit loss due to its overheating. In terms of power transmission operator who own transformer units, their loss can be covered under all-risk property policies, which is more general than name-peril policy. If utility has a name-peril policy, then the physical damage has to be caused by this specific peril. In terms of generation step-up transformers loss, the situation is more complicated. Generation step-up transformer loss leads to business interruption by preventing electricity sale. Even if the spare power transformer unit is stored at the power plant, it anyway takes up to several hours to change it (Table 4.2). Therefore, business interruption loss under the property cover would be claimed. In addition, legal actions by the stakeholders may come due to the companies underperformance. The manufactures of the failed transformers could face potential litigation from the owners/operators of the assets under product liability coverage part of a general commercial liability excess liability policy [156].

Torts against power companies for electricity undersupply resulting in property damage are not favorable for plaintiffs. Generally, other industries affected by GMD do not experience physical damage of assets. The loss originates from spilling perishable contents and/or operation/production ceasing. It opens a dialogue on how to treat the word "damage" as physical damage or the impairment of usefulness of the power companies. Claims after the 2003 US blackout are characteristic examples of how the definition "damage" can be threat by different parties. For instance, one of the restaurants made a claim under its insurance policy for food spoilage and loss of business income damages. At the end, it was approved by the court that an insurer has to limit the application of word "damage" in its policy to that which was "physical". Overall, tort actions for food spoilage and mould remediation are considered to be as the most successful ones [191].

There are two main forms of business interruption coverage as indicated by [192]: business interruption and contingent business interruption. Business interruption stays for the first party coverage for lost income due to physical damage at the insured locations. The example in the light of GMD impact is the generation step-up transformer loss. Contingent business interruption is used for third party coverage for lost income due to physical damage at suppliers, providers or consumers of its product or service. The other interconnected industries degradation or failure should be covered under this option. Service interruption cover is widely purchased by large corporations and covers outages of key utility service providers such as electricity, gas and water [193]. Homeowners who experience similar loss pattern on the smaller scale can be covered under HO-3 [194].

It is important to note that the physical damage to infrastructure happens on the transmission level. Therefore, the claim would be covered by the insurer in case a company has a direct contract with TSO. In case a company has a contract with a dispatching company, the insurance company may dispute the claim.

## 6.7 CONCLUSION

Extreme events, classified as catastrophes, are infrequent in their nature. It is not possible to identify the true behavior of extreme events from a limited set. Industry, risk managers and policy makers require the scientific community to give them information on:

- "how extreme future events can be?"
- "what is the expected frequency of such events?"
- "which damage can be expected?"

Natural forces can trigger hazardous events that affect modern technological systems. Electricity is the backbone of the modern economy, therefore its reliable operation is crucial. Entry points of the power system failure are determined by the type of the natural hazard. The GMD threat imposed on power grids is in the form of uncontrolled GICs flows over the network elements. It might result in some form of abnormal operation conditions or high likelihood of equipment damage. The unique feature of the power grid as a technological system is that the failure may occur in the region which is even not affected by the hazard.

GMDs can result in $k$ elements outage, where $k \geq 1$. The short forecast timescale makes it challenging to implement any meaningful preventive actions. Thereby, system vulnerabilities to GMDs should be defined in advance. The combination and analysis of relevant risk factors allow us to determine the power system vulnerability to GMDs.

It is shown that the technically predefined factors play a more important role than the naturally predefined ones. Among the technical factors which determine the power grid's robustness to GMDs, the most important ones are the construction scheme of the installed power system equipment and the power system topology and operation schemes. On the other hand, the awareness forms a self-standing group of factors (Fig. 6.17).

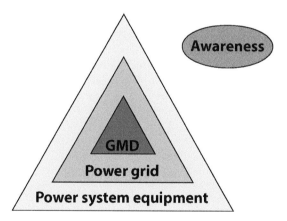

Figure 6.17: Hierarchy of the groups of critical factors that define power grid vulnerability to GMDs [138]

Industry is less concerned with reading specialized scientific reports and articles. The power grid hardening against GMD effects requires the development of tools, which are understandable to a wide audience (R2O). An example of such tools was presented in this chapter. The elaborated algorithm can be implemented on any power grid model. It can be practiced using the open source data by specialists without deep knowledge in geophysics. The proposed algorithm in this chapter allows us not only to identify the "high-risk" zones, but also it reveal the nodes that are particularly in danger. In other words, the algorithm highlights the places which are in the highest priority for the implementation of mitigation actions described in the previous chapter. The authors propose that the relevant stakeholders adopt the practice of preliminary investigation of power grid robustness to GMDs, even in the so-called mid-risk and low-risk regions.

## REFERENCES

1. Poljanšek, K., Bono, F., Gutiérrez, E. (2012). Seismic risk assessment of interdependent critical infrastructure systems: The case of European gas and electricity networks. *Earthquake Engineering & Structural Dynamics, 41(1)*, 61–79. doi: 10.1002/eqe.118.
2. Marzocchi, W., Garcia-Aristizabal, A., Gasparini, P., Mastellone, M. L., Di Ruocco, A. (2012). Basic principles of multi-risk assessment: a case study in Italy. *Natural Hazards, 62(2)*, 551–573. doi: 10.1007/s11069-012-0092-x.
3. Ward, D.M. (2013). The effect of weather on grid systems and the reliability of electricity supply. *Climatic Change*, 121:103–113.
4. SwissRe (2017). *Natural Catastrophes and Man-Made Disasters in 2016: A Year of Widespread Damages*. Sigma 2. Swiss Re, Switzerland.
5. Mukherjee, S., Nateghi, R., Hastak, M. (2018). A multi-hazard approach to assess severe weather-induced major power outage risks in the US. *Reliability Engineering & System*, 175:283–305.
6. Kenward, A., Urooj, R. (2014). Blackout: Extreme weather, climate change and power outages. *Climate Central*, 10:1–23.
7. Campbell, R.J. (2012). *Weather-Related Power Outages and Electric System Resiliency*. Congressional Research Service, Library of Congress Washington, DC.
8. ENTSO-E (2013). *Nordic grid disturbance statistics 2012*. ENTSO-E.
9. Rudnick, H. (2011). Impact of natural disasters on electricity supply. *IEEE Power and Energy Magazine*, 9:22–26.
10. Hoeppe, P. (2016). Trends in weather related disasters–Consequences for insurers and society. *Weather and Climate Extremes, 11*, 70–79. doi: 10.1016/j.wace.2015.10.002.
11. Mukhopadhyay, S., Nateghi, R. (2017). Estimating climate-Demand Nexus to support longterm adequacy planning in the energy sector. *2017 IEEE Power & Energy Society General Meeting*, 1–5.
12. Zamuda, C., Mignone, B., Bilello, D., et al. (2013). *US Energy Sector Vulnerabilities to Climate Change and Extreme Weather*. Department of Energy, Washington DC.
13. Voisin, N., Kintner-Meyer, N., Nguyen, T., et al. (2016). Vulnerability of the US western electric grid to hydro-climatological conditions: How bad can it get? *Energy*, 115:1–12.
14. North American Electric Reliability Corporation (NERC) (2012). *Severe impact resilience: Considerations and recommendations*. NERC, Atlanta, GA, USA.
15. World Economic Forum (2018). *The Global Risks Report 2018, 13th Edition*. World Economic Forum, Geneva.

16. Odenwald, S. (2010). Introduction to space storms and radiation. *Heliophysics: Space Storms and Radiation: Causes and Effects, 15*. Cambridge University Press, Cambridge.

17. Pulkkinen, A., Bernabeu, E., Thomson, A., et al. (2017). Geomagnetically induced currents: Science, engineering, and applications readiness. *Space Weather, 15(7)*, 828–856. doi: 10.1002/2016SW001501.

18. Love, J. J. (2012). Credible occurrence probabilities for extreme geophysical events: Earthquakes, volcanic eruptions, magnetic storms. *Geophysical Research Letters, 39(10)*. doi: 10.1029/2012GL051431.

19. Cannon, P. S. (2013). Extreme Space Weather–A report published by the UK Royal Academy of Engineering. *Space Weather, 11(4)*, 138–139. doi: 10.1002/swe.20032.

20. Schrijver, C.J., Mitchell, S.D. (2013). Disturbances in the US electric grid associated with geomagnetic activity. *Journal of Space Weather and Space Climate*, 3:A19.

21. Luntama, J.P. (2017). Report on the ESA Space-Weather Socio-Economic study. *Space Weather Manager ESA SSA Programme Office, Space Weather Workshop, 2–5 May 2017, Broomfield, Colorado*.

22. Menck, P. J., Heitzig, J., Kurths, J., Schellnhuber, H. J. (2014). How dead ends undermine power grid stability. *Nature communications, 5(1)*, 1–8. doi: 10.1038/ncomms4969

23. NERC (2012). *Interim Report: Effects of Geomagnetic Disturbances on the Bulk Power System*. North American Electricity Reliability Corporation.

24. Friedel, R.H.W., Reeves, G. D., Obara, T. (2002). Relativistic electron dynamics in the inner magnetosphere–A review. *Journal of Atmospheric and Solar-Terrestrial Physics, 64(2)*, 265–282. doi: 10.1016/S1364-6826(01)00088-8.

25. Pirjola, R. (2005). Averages of geomagnetically induced currents (GIC) in the Finnish 400kV electric power transmission system and the effect of neutral point reactors on GIC. *Journal of Atmospheric and Solar Terrestrial Physics*, 67:701–708.

26. Ngwira, C. M., Pulkkinen, A. A. (2018). An overview of science challenges pertaining to our understanding of extreme geomagnetically induced currents. In *Extreme Events in Geospace*, 187–208. Elsevier.

27. Onsager, T.G., Chan, A.A., Fei, Y., et al. (2004). The radial gradient of relativistic electrons at geosynchronous orbit. *Journal of Geophysical Research: Space Physics*, 109:A5.

28. Pulkkinen, A., Lindahl, S., Viljanen, A., Pirjola, R. (2005). Geomagnetic storm of 29–31 October 2003: Geomagnetically induced currents and their relation to problems in the Swedish high-voltage power transmission system. *Space Weather, 3(8)*. doi: 10.1029/2004SW00123

29. Viljanen, A., Tanskanen, E.I., Pulkkinen, A. (2006). Relation between substorm characteristics and rapid temporal variations of the ground magnetic field. *Annales Geophysicae*, 24:725–733.

30. Ngwira, C., Pulkkinen, A., Bernabeu, E., et al. (2015). Characteristics of extreme geoelectric fields and their possible causes: Localized peak enhancements. *Geophysical Research Letters*, 42:6916–6921.

31. Ngwira, C., Pulkkinen, A., Wilder, F., et al. (2013). Extended study of extreme geoelectric field event scenarios for geomagnetically induced current applications. *Space Weather*, 11:121–131.

32. Pulkkinen, A., Bernabeu, E., Eichner, J., Beggan, C., Thomson, A.W.P. (2012). Generation of 100-year geomagnetically induced current scenarios. *Space Weather, 10(4)*. doi: 10.1029/2011SW000750

33. Loeve, J., Coïsson, P., Pulkkinen, A. (2016). Global statistical maps of extreme-event magnetic observatory 1 min first differences in horizontal intensity. *Geophysical Research Letters*, 43:4126–4135.

34. Kappenman, J. (2006). Great geomagnetic storms and extreme impulsive geomagnetic field disturbance events–an analysis of observational evidence including the great storm of May 1921. *Advances in Space Research*, 38:188–199.

35. Fiori, R., Boteler, D., Gillies, D. (2014). Assessment of GIC risk due to geomagnetic sudden commencements and identification of the current systems responsible. *Space Weather*, 12:76–91.

36. Kappenman, J. G. (2003). Storm sudden commencement events and the associated geomagnetically induced current risks to ground-based systems at low-latitude and mid-latitude locations. *Space weather, 1(3)*. doi: 10.1029/2003SW00003.

37. Viljanen, A., Amm, O., Pirjola, R. (1999). Modeling geomagnetically induced currents during different ionospheric situations. *Journal of Geophysical Research: Space Physics, 104(A12)*, 28059–28071.

38. Pulkkinen, A., Kataoka, R. (2006). S-transform view of geomagnetically induced currents during geomagnetic superstorms. *Geophysical Research Letters, 33(12)*. doi: 10.1029/2006GL025822.

39. Kubota, Y., Kataoka, R., Den, M., Tanaka, T., Nagatsuma, T., Fujita, S. (2015). Global MHD simulation of magnetospheric response of preliminary impulse to large and sudden enhancement of the solar wind dynamic pressure. *Earth, Planets and Space, 67(1)*, 1–9. doi: 10.1186/s40623-015-0270-7.

40. Takeuchi, T., Araki, T., Viljanen, A., Watermann, J. (2002). Geomagnetic negative sudden impulses: Interplanetary causes and polarization distribution. *Journal of Geophysical Research: Space Physics, 107(A7)*, SMP-7. doi: 10.1029/2001JA900152.

41. Araki, T. (1994). A physical model of the geomagnetic sudden commencement. Solar wind sources of magnetospheric ultra-low-frequency waves. *Geophysical Monograph-American Geophysical Union*, 81:183–200.

42. Curto, J., Araki, T., Alberca, L. (2007). Evolution of the concept of sudden storm commencements and their operative identification. *Earth, Planets and Space*, 59:i–xii.

43. Elvidge, S., Angling, M. J. (2018). Using extreme value theory for determining the probability of Carrington-like solar flares. *Space Weather, 16(4)*, 417–421. doi: 10.1002/2017SW001727.

44. Silrergleit, V.M. (1996). On the occurrence of geomagnetic storms with sudden commencements. *Journal of Geomagnetism and Geoelectricity*, 48:1011–1016.

45. Tsubouchi, K., Omura, Y. (2007). Long-term occurrence probabilities of intense geomagnetic storm events. *Space Weather*, 5:1–12.

46. Silrergleit, V.M. (1999). Forecast of the most geomagnetically disturbed dayes. *Earth, Planets and Space*, 51:19–22.

47. Siscoe, GL. (1976). On the statistics of the largest geomagnetic storms per solar cycle. *Journal of Geophysical Research*, 81:4782–4784.

48. Thomson, A., Gaunt, C.T., Cilliers, P., et al. (2010). Present day challenges in understanding the geomagnetic hazard to national power grids. *Advances in Space Research*, 45:1182–1190.

49. Thomson, A., Dawson, E.B., Reay, S.J. (2011). Quantifying extreme behavior in geomagnetic activity. *Space Weather, 9(10)*. doi: 10.1029/2011SW000696

50. Ngwira, C.M., Pulkkinen, A., Kuznetsova, M.M., Glocer, A. (2014). Modeling extreme Carrington-type space weather events using three-dimensional global MHD simulations. *Journal of Geophysical Research: Space Physics*, 119:4456–4474.

51. Green, J.L., Boardsen, S. (2006). Duration and extent of the great auroral storm of 1859. *Advances in Space Research*, 38:130–135.
52. Silverman, S.M. (1995). Low latitude auroras: The storm of 25 September 1909. *Journal of Atmospheric and Terrestrial Physics*, 57:673–685.
53. Silverman, S.M., Cliver, E.W. (2001). Low-latitude auroras: the magnetic storm of 14–15 May 1921. *Journal of Atmospheric and Terrestrial Physics*, 63:523–535.
54. Silverman, S.M. (2008). Low-latitude auroras: The great aurora of 4 February 1872. *Journal of Atmospheric and Terrestrial Physics*, 70:1301–1308.
55. Schrijver, C.J., Dobbins, R., Murtagh, W., Petrinec, S.M. (2014). Assessing the impact of space weather on the electric power grid based on insurance claims for industrial electrical equipment. *Space Weather*, 12:487–498.
56. Riley, P., Baker, D., Liu, Y. D., Verronen, P., Singer, H., Güdel, M. (2018). Extreme space weather events: From cradle to grave. *Space Science Reviews, 214(1)*, 21. doi: 10.1007/s11214-017-0456-3.
57. Riley, P., Love, J.J. (2017). Extreme geomagnetic storms: Probabilistic forecasts and their uncertainties. *Space Weather*, 15:53–64.
58. Vasyliūnas, V.M. (2011). The largest imaginable magnetic storm. *Journal of Atmospheric and Solar-Terrestrial Physics*, 73:1444–1446.
59. Jonas, S., Fronczyk, K., Pratt, L.M. (2018). A framework to understand extreme space weather event probability. *Risk Analysis*, 38:1534–1540.
60. Riley, P. (2012). On the probability of occurrence of extreme space weather events. *Space Weather, 10(2)*, 1–12. doi: 10.1029/2011SW000734.
61. Kataoka, R. (2013). Probability of occurrence of extreme magnetic storms. *Space Weather, 11(5)*, 214–218. doi: 10.1002/swe.20044, 2013
62. Love, J.J., Rigler, E.J., Pulkkinen, A., Riley, P. (2015). On the lognormality of historical magnetic storm intensity statistics: Implications for extreme-event probabilities. *Geophysical Research Letters*, 42:6544–6553.
63. Lotz, S. I., Danskin, D. W. (2017). Extreme value analysis of induced geoelectric field in South Africa. *Space Weather, 15(10)*, 1347–1356. doi: 10.1002/2017SW001662
64. North American Reliability Corporation (2014). *Benchmark Geomagnetic Disturbance Event Description*. North American Reliability Corporation, Atlanta, GA
65. Bonner, L. R., Schultz, A. (2017). Rapid prediction of electric fields associated with geomagnetically induced currents in the presence of three-dimensional ground structure: Projection of remote magnetic observatory data through magnetotelluric impedance tensors. *Space Weather, 15(1)*, 204–227. doi: 10.1002/2016SW001535.
66. Kelbert, A., Balch, C.C., Pulkkinen, A., et al. (2016). Methodology for time-domain estimation of storm-time electric fields using the 3D Earth impedance. *Abstract GP23D-02, presented at 2016 Fall Meeting*, AGU San-Francisco, Calif., 12–16 Dec.
67. Weigel, R. S. (2017). A comparison of methods for estimating the geoelectric field. *Space Weather, 15(2)*, 430–440. doi: 10.1002/2016SW001504
68. Pulkkinen, A., Amm, O., Viljanen, A. (2003). Ionospheric equivalent current distributions determined with the method of spherical elementary current systems. *Journal of Geophysical Research: Space Physics, 108(A2)*. doi: 10.1029/2001JA005085.
69. Rigler, E.J., Pulkkinen, A.A., Balch, C.C., et al. (2014). Dynamic geomagnetic hazard maps in space weather operations. *Abstract SM31A-4178, presented at 2014 Fall Meeting, AGU San-Francisco, Calif., 15-19 Dec.*
70. Love, J. J., Bedrosian, P. A., Schultz, A. (2017). Down to Earth with an electric hazard from space. *Space Weather, 15(5)*, 658–662. doi: 10.1002/2017SW001622.

71. Kappenman, J. (2010). *Geomagnetic Storms and Their Impacts on the US Power Grid.* Metatech, CA.

72. Lundstedt, H. (2006). The sun, space weather and GIC effects in Sweden. *Advances in Space Research*, 37:1182–1191.

73. ENTSO-E (2014). *Continental Europe Operation Handbook.* UCTE.

74. Piccinelli, R., Krausmann, E. (2018). North Europe power transmission system vulnerability during extreme space weather. *Journal of Space Weather and Space Climate, 8,* A03. doi: 10.1051/swsc/2017033.

75. Viljanen, A., Pirjola, R., Prácser, E., et al. (2013). Geomagnetically induced currents in Europe: characteristics based on a local power grid model. *Space Weather*, 11:575–584.

76. Viljanen, A., Pirjola, R., Wik, M., Ádám, A., Prácser, E., Sakharov, Y., Katkalov, J. (2012). Continental scale modelling of geomagnetically induced currents. *Journal of Space Weather and Space Climate, 2,* A17. doi: 10.1051/swsc/2012017.

77. Ádám, A., Prácser, E., Wesztergom, V. (2012). Estimation of the electric resistivity distribution (EURHOM) in the European lithosphere in the frame of the EURISGIC WP2 project. *Acta Geodaetica et Geophysica Hungarica*, 47:377–387.

78. Pulkkinen, A., Pirjola, R., Viljanen, A. (2008). Statistics of extreme geomagnetically induced current events. *Space Weather, 6(7)*, 1–10. doi: 10.1029/2008SW000388.

79. Pulkkinen, A., Viljanen, A., Pirjola, R. (2006). Estimation of geomagnetically induced current levels from different input data. *Space Weather, 4(8)*. doi: 10.1029/2006SW000229.

80. Pirjola, R. (2008). Effects of interactions between stations on the calculation of geomagnetically induced currents in an electric power transmission system. *Earth, Planets and Space*, 60:743–751.

81. Myllys, M., Viljanen, A., Rui, O., et al. (2014). Geomagnetically induced currents in Norway: the northernmost high-voltage power grid in the world. *Journal of Space Weather and Space Climatec*, 4:A10.

82. Pirjola, R. J., Viljanen, A. T., Pulkkineni, A. A. (2007). Research of geomagnetically induced currents (GIC) in Finland. In *2007 7th International Symposium on Electromagnetic Compatibility and Electromagnetic Ecology*, 269–272. IEEE.

83. Boteler, D.H. (2013). The use of linear superposition in modelling geomagnetically induced currents. In *2013 IEEE Power & Energy Society General Meeting*, 1–5.

84. Zheng, K., Boteler, D., Pirjola, R., et al. (2013). Effects of system characteristics on geomagnetically induced currents. *IEEE Transactions on Power Delivery*, 29:890–898.

85. Caraballo, R. (2016). Geomagnetically induced currents in Uruguay: Sensitivity to modelling parameters. *Advances in Space Research, 58(10)*, 2067–2075.

86. Pirjola, R., Liu, C. M., Liu, L. G. (2010). Geomagnetically induced currents in electric power transmission networks at different latitudes. In *2010 Asia-Pacific International Symposium on Electromagnetic Compatibility*, 699–702. IEEE. doi: 10.1109/APEMC.2010.5475727.

87. Space Studies Board (2008). *Severe Space Weather Events: Understanding Societal and Economic Impacts.* National Academy Press, Washington, D. C.

88. Aytac, A., Atay, B. (2015). A measure of global efficiency in networks. *International Journal of Pure and Applied Mathematics*, 103:61–70.

89. Arajärvi, E., Pirjola, R., Viljanen, A. (2011). Effects of neutral point reactors and series capacitors on geomagnetically induced currents in a high-voltage electric power transmission system. *Space Weather, 9(11)*, 1–10. doi: 10.1029/2011SW000715.

90. Volkov, A. I., Korovkin, N. V., Sokolova, O. N., Sorokin, E. V., Frolov, O. V. (2011). Method for optimizing control actions following emergencies in large-city electric power systems. *Power Technology and Engineering, 45(1)*, 50–52. doi: 10.1007/s10749-011-0222-8.

91. Finucane, M.L., Alhakami, A., Slovic, P., et al. (2000). The affect heuristic in judgments of risks and benefits. *Journal of Behavioral Decision Making*, 13:61–17.

92. Xie, X., Wang, M., Xu, L. (2003). What risks are Chinese people concerned about? *Risk Analysis: An International Journal*, 23:685–695.

93. Mann, L., Burnett, P., Radford, M., et al. (1997). The Melbourne Decision Making Questionnaire: An instrument for measuring patterns for coping with decisional conflict. *Journal of Behavioral Decision Making*, 10:1–19.

94. Janis, I.L, Mann, L. (1977). *Decision making: A psychological analysis of conflict, choice, and commitment*. Free Press, Washington D.C.

95. Brown, J., Abdallah, S.S., Ng, R. (2011). Decision making styles East and West: Is it time to move beyond cross-cultural research?. *International Journal of Sociology and Anthropology*, 3:452–459.

96. Di Fabio, A. (2006). Decisional procrastination correlates: Personality traits, self-esteem or perception of cognitive failure?. *International Journal for Educational and Vocational Guidance*, 6:109–122.

97. Boardman, J., Sauser, B. (2008). *Systems Thinking: Coping with 21st Century Problems*. CRC Press, Boca Raton FL.

98. Hall, M. L. (1990). Chronology of the principal scientific and governmental actions leading up to the November 13, 1985 eruption of Nevado del Ruiz, Colombia. *Journal of Volcanology and Geothermal Research, 42(1-2)*, 101–115. doi: 10.1016/0377-0273(90)90072-N.

99. Nejedlik, P., Dalezios, N.R. (2017). Hazards information management and services. *Environmental Hazards: Methodologies for Risk Assessment and Management*, 503. IWA Publishing.

100. Goodman, R.M., Speers, M.A., McLeroy, K., et al. (1998). Identifying and defining the dimensions of community capacity to provide a basis for measurement. *Health Education & Behavior*, 25:258–278.

101. Norris, F.H., Stevens, S.P., Pfefferbaum, B., et al. (2008). Community resilience as a metaphor, theory, set of capacities, and strategy for disaster readiness. *American Journal of Community Psychology*, 41:127–150.

102. Lee, B., Preston, F., Green, G. (2012). *Preparing for High-Impact, Low-Probability Events: Lessons from Eyjafjallajökull*. Chatham House, London.

103. Lazo, J.K., Kinnell, J.S., Fisher, A. (2000). Expert and layperson perceptions of ecosystem risk. *Risk Analysis*, 20:179–194.

104. Siegrist, M., Gutscher, H. (2006). Flooding risks: A comparison of lay people's perceptions and expert's assessments in Switzerland. *Risk Analysis*, 25:971–979.

105. Rowe, G., Wright, G. (2001). Differences in expert and lay judgments of risk: myth or reality?. *Risk Analysis*, 21:341–356.

106. United Nations Committee on Peaceful Uses of Outer Space (UNOOSA). (2017). *United Nations/United Arab Emirates High Level Forum: Space as a driver for socio-economic sustainable development*. United Arab Emirates Space Agency, Dubai.

107. Alexander, D. (2013). Volcanic ash in the atmosphere and risks for civil aviation: A study in European crisis management. *International Journal of Disaster Risk Science, 4(1)*, 9–19. doi: 10.1007/s13753-013-0003-0.

108. Mohrbach, L., PowerTech eV, V.G.B. (2013). Fukushima two years after the tsunami–the consequences worldwide. *Atomwirtschaft, 58(3)*, 152.

109. Holub, M., Fuchs, S. (2009). Mitigating mountain hazards in Austria – Legislation, risk transfer, and awareness building. *Natural Hazards and Earth System Sciences*, 9:523–527.

110. King, D. (2000). You are on your own: Community vulnerability and the need for awareness and education for predictable natural disasters. *Journal of Contingencies and Crisis Management*, 8:223–228.

111. Crosweller, H. S., Wilmshurst, J. (2013). *Natural Hazards and Risk: the Human Perspective. Risk and Uncertainty Assessment for Natural Hazards*. Cambridge University Press: Cambridge, 548–569.

112. Paton, D., Sagala, S., Okada, N., et al. (2010). Making sense of natural hazard mitigation: Personal, social and cultural influences. *Environmental Hazards*, 9:183–196.

113. Weyrich, P., Scolobig, A., Bresch, D.N., et al. (2018). Effects of impact-based warnings and behavioral recommendations for extreme weather events. *Weather, Climate, and Society*, 10:781–796.

114. Fearnley, C., Winson, A.E.G., Pallister, J., Tilling, R. (2017). Volcano crisis communication: challenges and solutions in the 21st century. *Observing the Volcano World*, 3–21. Springer, Cham.

115. Funabashi, Y., Kitazawa, K. (2008). Fukushima in review: A complex disaster, a disastrous response. *Bulletin of the Atomic Scientists*, 68:9–21.

116. Hatamura, Y., Oike, K., Kakinuma, T., et al. (2011). *Executive Summary of the Interim Report*. Investigation Committee on the Accident at Fukushima Nuclear Power Stations of Tokyo Electric Power Company.

117. Thompson, K.D., Stedinger, J.R., Heath, D.C. (1997). Evaluation and presentation of dam failure and flood risks. *Journal of Water Resources Planning and Management*, 123:216–227.

118. Lanabere, V., Dasso, S., Gulisano, A. M., López, V. E., Niemelä-Celeda, A. E. (2020). Space weather service activities and initiatives at LAMP (Argentinean Space Weather Laboratory group). *Advances in Space Research, 65(9)*, 2223–2234. doi: 10.1016/j.asr.2019.08.016.

119. Denardini, C.M., Dasso, S., Gonzalez-Esparza, J.A. (2016). Review on space weather in Latin America. 1. The beginning from space science research. *Advances in Space Research*, 58:1916–1939.

120. Denardini, C.M., Dasso, S., Gonzalez-Esparza, J.A. (2016). Review on space weather in Latin America. 2. The research networks ready for space weather. *Advances in Space Research*, 58:1940–1959.

121. Gonzalez-Esparza, J.A., De la Luz, V., Corona-Romero, P., et al. (2017). Mexican space weather service (SCIESMEX). *Space Weather*, 15:3–11.

122. Denardini, C.M., Chen, S.S., Resende, L.C.A., et al. (2018). The embrace magnetometer network for South America: Network description and its qualification. *Radio Science*, 53:288–302.

123. Elovaara, J. (2007). Finnish experiences with grid effects of GIC's. *Space Weather*, 311–326. Springer, Dordrecht.

124. Wik, M., Pirjola, R., Lundstedt, H., Viljanen, A., Wintoft, P., Pulkkinen, A. (2009). Space weather events in July 1982 and October 2003 and the effects of geomagnetically induced currents on Swedish technical systems. *Annales Geophysicae, 27(4)*, 1775–1787. Copernicus GmbH.

125. Barannik, M., Danilin, A., Katkalov, Yu., et al. (2012). A system for recording geomagnetically induced currents in neutrals of power autotransformers. *Instruments and Experimental Techniques*, 55:110–115.

126. Torta, J. M., Serrano, L., Regué, J. R., Sanchez, A. M., Roldán, E. (2012). Geomagnetically induced currents in a power grid of northeastern Spain. *Space Weather, 10(6)*. doi: 10.1029/2012SW000793.

127. Thomson, A. W., McKay, A. J., Clarke, E., Reay, S. J. (2005). Surface electric fields and geomagnetically induced currents in the Scottish Power grid during the 30 October 2003 geomagnetic storm. *Space Weather, 3(11)*. doi: 10.1029/2005SW000156

128. Erinmez, I. A., Kappenman, J. G., Radasky, W. A. (2002). Management of the geomagnetically induced current risks on the National grid company's electric power transmission system. *Journal of Atmospheric and Solar-Terrestrial Physics*, 64(5-6), 743–756.

129. Kelly, G. S., Viljanen, A., Beggan, C. D., Thomson, A. W. P. (2017). Understanding GIC in the UK and French high-voltage transmission systems during severe magnetic storms. *Space Weather*, 15(1), 99–114.

130. Tozzi, R., De Michelis, P., Coco, I., Giannattasio, F. (2019). A preliminary risk assessment of geomagnetically induced currents over the Italian territory. *Space Weather, 17(1)*, 46–58. doi: 10.1029/2018SW002065

131. Carter, B. A., Yizengaw, E., Pradipta, R., Halford, A. J., Norman, R., Zhang, K. (2015). Interplanetary shocks and the resulting geomagnetically induced currents at the equator. *Geophysical Research Letters*, 42(16), 6554–6559.

132. Wei, L.-H., Homeier, N., Gannon, J. L. (2013). Surface electric fields for North America during historical geomagnetic storms. *Space Weather*, 11(8), 451–462.

133. Love, J.-J., Joshua Rigler, E., Pulkkinen, A., Balch, C. C. (2014). Magnetic storms and induction hazards. *EOS, Transactions American Geophysical Union*, 95(48), 445–446. doi: 10.1002/2014EO480001.

134. Love, J.-J., Pulkkinen, A., Bedrosian, P. A., et al. (2016). Geoelectric hazard maps for the continental United States. *Geophysical Research Letters*, 43(18), 9415–9424.

135. Love, J. J., Lucas, G. M., Kelbert, A., Bedrosian, P.-A. (2018). Geoelectric hazard maps for the Pacific Northwest. *Space Weather, 16(8)*, 1114–1127.

136. Viljanen, A. (2011). European project to improve models of geomagnetically induced currents. *Space Weather, 9(7)*. doi: 10.1029/2011SW000680

137. Viljanen, A., Pirjola, R., Prácser, E., Katkalov, J., Wik, M. (2014). Geomagnetically induced currents in Europe-Modelled occurrence in a continent-wide power grid. *Journal of Space Weather and Space Climate, 4*, A09. doi: 10.1051/swsc/2014006.

138. Sokolova, O., Burgherr, P., Sakharov, Ya., Korovkin, N. (2018). Algorithm for analysis of power grid vulnerability to geomagnetic disturbances. *Space Weather*, 16(10), 1570–1582. doi: 10.1029/2018SW001931.

139. Sokolova, O., Popov, V. (2017). Critical infrastructure exposure to severe solar storms. Case of Russia. In *Safety and Reliability-Theory and Applications-Proceedings of the 27th European Safety and Reliability Conference, ESREL 2017*, 1327–1334.

140. International Energy Agency (2013). *Key World Energy Statistics*, 6. International Energy Agency, Paris, France.

141. International Energy Agency (2018). *Global Energy & CO2 Status Report. The latest trends in energy and emissions in 2018*. International Energy Agency, Paris, France.

142. MacAlester, M.-H., Murtagh, W. (2014). Extreme space weather impact: An emergency management perspective. *Space Weather*, 12(8), 530–537.

143. Zio, E. (2016). Critical infrastructures vulnerability and risk analysis. *European Journal for Security Research, 1(2)*, 97–114.

144. Rinaldi, S.-M., Peerenboom, J.-P., Kelly, T.-K. (2001). Identifying, understanding, and analyzing critical infrastructure interdependencies. *IEEE Control Systems Magazine, 21(6)*, 11–25.

145. Foster, Jr., John, S., Gjelde, E., et al. (2008). *Report of the Commission to Assess the Threat to the United States from Electromagnetic Pulse (EMP) Attack: Critical national infrastructures*. Defense Technical Information Center, USA.

146. Koks, E., Pant, R., Thacker, S., Hall, J. W. (2019). Understanding business disruption and economic losses due to electricity failures and flooding. *International Journal of Disaster Risk Science, 10(4)*, 421–438. doi: 10.1007/s13753-019-00236-y.

147. Petermann, T., Bradke, H., Lüllmann, A., Poetzsch, M., Riehm, U. (2014). *What Happens During a Blackout: Consequences of a Prolonged and Wide-Ranging Power Outage*. Office of Technology Assessment at the German Bundestag.

148. Klein, K.-R., Rosenthal, M.-S., Klausner, H.-A. (2005). Blackout 2003: preparedness and lessons learned from the perspectives of four hospitals. *Prehospital and Disaster Medicine, 20(5)*, 343.

149. Wasson, R. J. (2018). Zaps and taps: solar storms, electricity and water supply disasters, and governance. *Crossing Borders*, 261–277. Springer, Singapore.

150. Forbes, K.-F., Cyr, O.-S. (2004). Space weather and the electricity market: An initial assessment. *Space Weather*, 2(10), 1–28. doi: 10.1029/2003SW000005.

151. Forbes, K.-F., Cyr, O.-S. (2008). Solar activity and economic fundamentals: Evidence from 12 geographically disparate power grids. *Space Weather*, 6(10), 1–20. doi: 10.1029/2007SW000350

152. Forbes, K. F., St. Cyr, O. C. (2010). An anatomy of space weather's electricity market impact: Case of the PJM power grid and the performance of its 500 kV transformers. *Space Weather, 8(9)*. doi: 10.1029/2009SW000498

153. Uritskaya, O., Robinson, R.M. (2019). *Direct Economic Evaluation of Impact of Severe Space Weather Events on a Deregulated Electricity Market*. AGU Fall Meeting 2019.

154. Weare, C. (2003). *The California Electricity Crisis: Causes and Policy Options*. Public Policy Institute, San Francisco.

155. Munasinghe, M., Sanghvi, A. (1998). Reliability of electricity supply, outage costs and value of service: An overview. *The Energy Journal*, 9:1–18

156. Oughton, E., Copic, J., Skelton, A., et al. (2016). *Helios Solar Storm Scenario. Cambridge Risk Framework Series*. Centre for Risk Studies, University of Cambridge.

157. Kelly, S. (2015). Estimating economic loss from cascading infrastructure failure: a perspective on modelling interdependency. *Infrastructure Complexity*, 2(1), 7. doi: 10.1186/s40551-015-0010-y.

158. Timmer, M., Los, B., Stehrer, R., de Vries, G. (2016). *An Anatomy of the Global Trade Slowdown Based on the WIOD 2016 release*. Groningen Growth and Development Centre, University of Groningen.

159. Royal Academy of Engineering (2014). *Counting the Cost: the Economic and Social Costs of Electricity Shortfalls in the UK*. Royal Academy of Engineering, London.

160. de Nooij, M., Koopmans, C., Bijvoet, C. (2003). The demand for supply security. *Research Symposium European Electricity Markets*. The Hague, September 2003.

161. Bolduc, L. (2002). GIC observations and studies in the Hydro-Québec power system. *Journal of Atmospheric and Solar-Terrestrial Physics*, 64(16), 1793–1802. doi: 10.1016/S1364-6826(02)00128-1.

162. Spence, S., Morley, K. M. (2012). A lesson in resilience from Derecho. *Journal-American Water Works Association, 104(9)*, 20–23.
163. Eastwood, J. P., Hapgood, M. A., Biffis, E., Benedetti, D., Bisi, M. M., Green, L., et al. (2018). Quantifying the economic value of space weather forecasting for power grids: An exploratory study. *Space Weather*, 16(12), 2052–2067. doi: 10.1029/2018SW002003
164. Moran, D., Lenzen, M., Cairns, I. H., Steenge, A. E. (2014). How severe space weather can disrupt global supply chains?. *Natural hazards Earth Systems Science*, 14:2749–2759.
165. Maynard, T., Smith, N., Gonsalez, S. (2013). *Solar Storm Risk to the North American Electric Grid.* Lloyd's, London.
166. MacAlester, M.-H. (2018). Extreme space weather and emergency management. *Extreme Events in Geospace*, 683–700. Elsevier.
167. Schrijver, C. J., Kauristie, K., Aylward, A. D., Denardini, C. M., Gibson, S. E., Glover, A., Jakowski, N. (2015). Understanding space weather to shield society: A global road map for 2015—2025 commissioned by COSPAR and ILWS. *Advances in Space Research*, 55(12), 2745–2807.
168. McMorrow, D. (2011). Impacts of severe space weather on the electric grid. *JASON*, 22102–7508. Virginia.
169. United States Government Accountability Office. (2018). *GAO-19-98 Technology assessment. Critical infrastructure protection. Protecting the electric grid from geomagnetic disturbances.* United States.
170. Jonkeren, O., Azzini, I., Galbusera, L., Ntalampiras, S., Giannopoulos, G. (2015). Analysis of critical infrastructure network failure in the European Union: A combined systems engineering and economic model. *Networks and Spatial Economics*, 15(2), 253–270. doi: 10.1007/s11067-014-9259-1.
171. Abedi, A., Gaudard, L., Romerio, F. (2019). Review of major approaches to analyze vulnerability in power system. *Reliability Engineering & System Safety*, 183, 153–172.
172. Carrera, L., Standardi, G., Bosello, F., Mysiak, J. (2015). Assessing direct and indirect economic impacts of a flood event through the integration of spatial and computable general equilibrium modelling. *Environmental Modelling & Software*, 63, 109–122. doi: 10.1016/j.envsoft.2014.09.016.
173. Kahn, M.-E. (2005). The death toll from natural disasters: the role of income, geography, and institutions. *Review of Economics and Statistics*, 87(2), 271–284.
174. Toya, H., Skidmore, M. (2007). Economic development and the impacts of natural disasters. *Economics Letters*, 94(1), 20–25.
175. Felbermayr, G., Gröschl, J. (2014). Naturally negative: The growth effects of natural disasters. *Journal of Development Economics*, 111, 92–106. doi: 10.1016/j.jdeveco.2014.07.004.
176. Cavallo, E., Noy, I. (2011). Natural disasters and the economy—a survey. *International Review of Environmental and Resource Economics*, 5(1), 63–102.
177. Hapgood, M., Thomson, A. (2010). *Space Weather: Its Impact on Earth and Implications for Business.* Lloyd's 360 Risk Insight, London.
178. Skoufias, E. (2003). Economic crises and natural disasters: Coping strategies and policy implications. *World development*, 31(7), 1087–1102.
179. Schrijver, C. J. (2015). Socio-economic hazards and impacts of space weather: The important range between mild and extreme. *Space Weather*, 13(9), 524–528., doi: 10.1002/2015SW001252.

180. Borscheid, P., Gugerli, D., Straumann, T. (2013). *The Value of Risk: Swiss Re and the History of Reinsurance*. OUP Oxford.

181. Kaplan, S., Garrick, B. J. (1981). On the quantitative definition of risk. *Risk Analysis*, 1(1), 11–27.

182. Lloyd's Market Association (LMA) (2013). *Catastrophe Modelling: Guidance for Non-Catastrophe Modelers*. Lloyd's Market Association, London.

183. Grossi, P. (2005). Catastrophe modeling: a new approach to managing risk. *Springer Science & Business Media*.

184. Mitchell-Wallace, K., Jones, M., Hillier, J., Foote, M. (2017). *Natural Catastrophe Risk Management and Modelling: A Practitioner's Guide*. John Wiley & Sons, Croydon.

185. Gunasekera, R., Ishizawa, O., Aubrecht, C., Blankespoor, B., Murray, S., Pomonis, A., Daniell, J. (2015). Developing an adaptive global exposure model to support the generation of country disaster risk profiles. *Earth-Science Reviews, 150*, 594–608. doi: 10.1016/j.earscirev.2015.08.012.

186. Rossetto, T., Ioannou, I., Grant, D. N. (2013). *Existing Empirical Fragility and Vulnerability Relationships: Compendium and Guide for Selection*. GEM Foundation, Pavia.

187. Bank of England, Prudential regulation authority. (2019). *General Insurance Stress Test 2019. Scenario Specification, Guidelines and Instructions*. Prudential Regulation Authority, London.

188. Sokolova, O., Burgherr, P., Collenberg, W., Schwerzmann, A. (2014). *The Impact of Solar Storms on Power Systems*. Swiss Re, Zurich.

189. AON Benfield (2013). *Geomagnetic storms*. AON Benfield, Minneapolis, MN.

190. Dobbins, R.W., Schriiver, K. (2014). *Electrical Claims and Space Weather. Measuring the Visible Effects of an Invisible Force*. Zurich Insurance Group Ltd., Zurich.

191. Standler, R. B. (2011). *Liability of electric utility in the USA for outage or blackout*, 51. Accessible on `http://www. rbs2. com/outage. pdf` (Accessed 15 June 2020).

192. Berry, D. (2000). *The basics of a business interruption claim*. Accessible on `http://www.irmi.com/articles/expert-commentary/ the-basics-of-a-business-interruption-claim.` (Accessed 15 June 2020).

193. Launay, R. (2014). *Solar Storms and Their Impacts on Power Grids – Recommendations for(re)insurers*. ISCPR Papers, no.28.

194. Agrella, R. (2015). *Does homeowners insurance cover solar storm damage?*. Insuramatch, March 19.

# APPENDIX

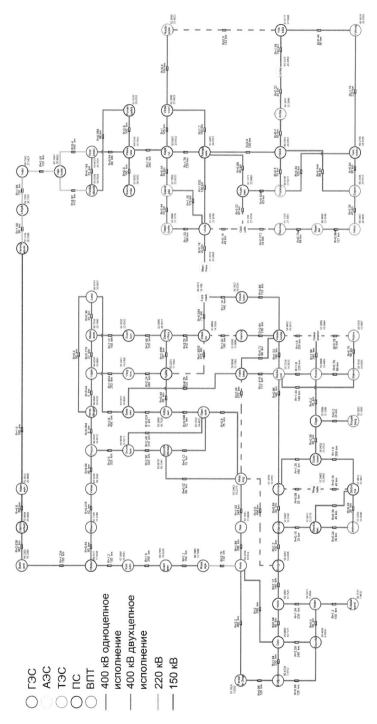

Figure 6.18: Equivalent 400 kV power system of Scandinavia

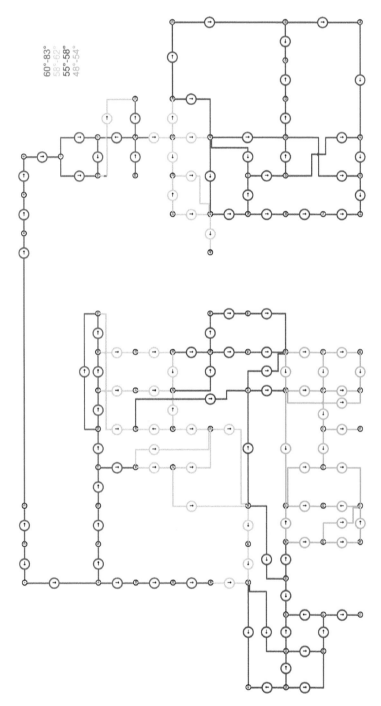

Figure 6.19: Scandinavian power system according to geomagnetic latitudes distribution

# 7 Outlook

"AN OUNCE OF PREVENTION IS WORTH A POUND OF CURE".

**- Benjamin Franklin**
*1706-1790*

STRONG GMDs will hit the Earth again. It is not a matter of *if*, but *when*, and, more general, *how much will we on Earth be vulnerable to GMD effects at that moment?* The OECD considers the GMD as a threat on the global shock level together with pandemics, financial crises and cyber risks. These types of events are all caused by an external force, simultaneously spread over a large geographic area, and potentially result in enormous economic losses. The modern history shows that hazard has no borders, though risk does, due to national and regional differences in vulnerabilities and exposure landscapes.

Power systems have a long lifetime; thus one should think of a heightened awareness of rare events. Dealing with the risk of a 100-year event is not a day-to-day business of grid owners and operators. However, day-to-day challenges and routines should not prevent operators from developing a complete spectrum of mitigation activities. One can compare GMDs with ghosts, living in a house! It is not possible to visualize them, as they existed for much longer period than one was cautious about their existence and impact; they act with no or little precaution. Detailed studies show to the society that a "zero risk level" region does not exist.

In this book, a comprehensive review on the current state of understanding GMD hazard is presented. Looking into the future, the increase of global population to 9 billion or perhaps beyond, and the electricity growth rate at 2.1% per year will escalate even more the GMD risk. Prevention is the poor man in the disaster management. Compared to response and recovery, it stays in the shadow. The aim of the comprehensive analysis in this book is to help various stakeholders including disaster managers, governments, power grid owners and operators to be better prepared for the uncertain future GMD events.

The risk governance of GMD impact on power grids should be viewed as an interdisciplinary problem. This is the concept taken in this book, as vulnerabilities are studied in relation to the internal and external influences. The common measures of "how hazardous is a GMD" are the $I_{GIC}$ amplitude and the rate-of-change $dB/dt$.

They are measurable, interpretable, and accessible. They track and reflect variabilities in the hazard development. As it is shown in Chapter 3, the established observation methods are robust. However, accurate correlation of $I_{GIC}$ amplitude with the level of damage is still an open question in the GMD risk governance.

The attempt of multicriterial analysis of power grid robustness to GMDs is presented in this book. The list of critical factors is proposed, described and ranked in accordance with their level of impacts. It is discussed that similar events can have quite different impacts and consequences. Therefore, other measures apart from $I_{GIC}$ and $dB/dt$ play a role. Industry is less concerned with reading specialized scientific reports and articles. The power grid hardening against GMD effects requires the developments of tools which are understandable to a wide audience. An example of such tools is presented here. The evaluation of risks provides a basis for planning and allocating limited resources: technical, financial, and others.

In contrast to other natural hazards, risk transfer mechanisms for GMDs are less developed. Risk can be defined as the set of a given scenario, which is the basis of catastrophe models. Based on the scenario, the amount of risk to be transferred and the price for it are specified. The GMD, which serves as a benchmark for a worst-case scenario (the Carrington event), happened at a time when technological systems were relatively undeveloped. According to GMD metrics, only one true extreme event happened in the modern space era (Hydro-Québec blackout). As power grid susceptibility to GMD effects is a multicriterial problem, the accurate estimation of, how worst scenario can be, is our main concern. Nevertheless, mitigation procedures should be developed for the whole multitude of events. Further research may be focused on enriching the knowledge of the complete GMD event chain, the development of more accurate fragility curves together with the improved modeling of vulnerabilities.

The hazard assessment is important for designing mitigation schemes, which are chosen in an ever-lasting problem of maximizing the profits and compromising the safety. The impact of severe GMDs can cross geographical/political frontiers. In other words, the crisis in one country can affect infrastructure operation in the neighboring countries. The risk management paradigm shifts towards proactive actions for improving overall system's resilience. The concept of resilience extends the traditional system limits.

In this book, we suggest to classify an extreme GMD not as a high-impact low-frequency event, but as a perfect storm. Accepting this assumption significantly changes the mitigation landscape opportunities. It is shown that ultimately extreme events are better predictable, since more time is required for their development in the Sun's corona. Therefore, the main matter requiring resolution is "what is the strongest storm with low predictability, which is hazardous for the modern technological systems?". The GMD as a hazard does not have a unique way to endanger power grid operation, contrary to other natural hazards. The grid operators and planners have to keep in mind three failure scenarios described in Chapter 4. Furthermore, the actions and procedures toward an increased resiliency described in Chapter 5 are doable and essential.

# Glossary

**Aggregate exceedance probability (AEP):** The probability of the sum of event losses in a year exceeding a certain level.

**Aurora:** A natural display of light in the polar sky produced by the collision of energetic electrons with atoms and molecules of gases such as oxygen and nitrogen in the upper ionosphere.

**Auroral oval:** A band of latitudes in both hemispheres where aurora appears.

**Average annual loss (AAL):** The expected loss cost over a one-year time period.

**Benchmark:** Model that serves as a standard.

**Bus:** A conductor, or group of conductors that serves as a common connection for two or more circuits.

**Business interruption (BI):** A coverage category that provides protection against lost profit or revenue due to damage, which results in the business not being able to properly function.

**Catastrophe:** An event that exceeds capability of those affected to cope with, or absorb its effects.

**Circuit Breaker:** A mechanical switching device, capable of making, carrying, and breaking currents under normal circuit conditions and also, making and carrying for a specified time and breaking currents under specified abnormal circuit conditions such as those of short circuit.

**Contingency:** The unexpected failure or outage of a system component, such as a generator, transmission line, circuit breaker, switch or other electrical element.

**Corona:** The outermost layer of the solar atmosphere, characterized by low densities in the range of $(\leq 10^9)$ cm$^{-3}$ – $(10^{15})$ m$^{-3}$ and high temperatures $(\geq 10^6)$ K

**Coronal hole:** Solar corona region characterized by exceptionally low density and in a unipolat photospheric magnetic field having open magnetic field lines.

**Coronal mass ejection:** Ejections of high-mass clouds of solar material in the Sun's corona into the heliosphere. They are mainly associated with disappearing solar filaments, erupting prominences, and solar flares.

**Cost/risk criteria:** Criteria of the loss and required cost for a system reinforcement.

**Corotating interaction region:** Region created when a high speed solar wind stream emanating from a coronal hole overtakes a slower (upstream) solar wind stream.

**Damage ratio:** The estimated repair cost of an asset at risk divided by the replacement cost of an asset.

**Dipole field:** The arrangement of two opposite magnetic poles of equal strength (North and South poles). The pattern of the magnetic field lines in permanent magnets is a dipole field.

**Disaster:** A serious disruption of the functioning of a community or a society at any scale due to hazardous events interacting with conditions of exposure, vulnerability and capacity, leading to one or more of the following: human, material, economic and environmental losses and impacts.

**Dst index:** The disturbance storm index is a measure of the horizontal magnetic field variations due to the presence of the enhanced equatorial ring current.

**Ecliptic plane:** The plane of the Earth's orbit about the Sun.

**Electrojet:** (1) Auroral Electrojet (AE): the current that flows in the ionosphere at a height of $\sim$100 km in the auroral zone. (2) Equatorial Electrojet: the thin electric current layer in the ionosphere over the dip equator at about 100 to 115 km altitude.

**Emerging risk:** An issue that is perceived to be potentially significant but which may not be fully understood.

**Ensemble forecasts:** Probabilistic forecasts defining the spectrum of what could happen, when using a number of different models.

**Expected loss:** The statistical mean loss of a quantity.

**Exposure data:** The data representing the assets to be modeled.

**Extreme value analysis:** A branch of statistics dealing with the extreme deviations from the median of probability distributions.

**Flare:** The outbursts in the chromosphere of the Sub, during which a limited area rapidly and temporarily becomes very bright.

**Galactic cosmic rays:** Particle radiation originating from the stars in the Milky Way (our galaxy).

**Geographical information system (GIS):** A digitalized system for capturing, storing, checking, and displaying data related to positions on Earth's surface.

**Geomagnetic disturbance:** A worldwide disturbance of the Earth's magnetic field, distinct from regular diurnal variations. A storm is precisely defined as occurring when Dst index becomes less than $-50$ nT.

**Geomagnetically induced currents:** Ultra-low frequency electric currents flowing within a technical system induced by electric fields caused by a geomagnetic field change over time.

**Governance:** The processes, controls and oversight put in place for ensuring that catastrophe risk is properly managed.

**Grid instability:** A phenomenon, typically with respect to voltage, that causes an electric grid to fail due to collapsing voltage that propagates across the grid.

**Hazard:** A process, phenomenon or human activity that may cause loss of life, injury or other health impacts, property damage, social and economic disruption or environmental degradation. GMD is considered as a hazard and blackout caused by GMD is a disaster.

**Heliosphere:** Space around Sun defined as the area, in which the pressures of the solar wind and the interstellar medium are equal.

**High voltage:** Voltage levels higher than $\geq 110$ kV

**Indirect loss:** Loss caused by the wider impacts of an event.

**Insurance:** An arrangement whereby one party (insurer) promises to pay another party (the policyholder) a sum of money in the event of a loss due to a specific case.

**Interplanetary coronal mass ejection:** Type of coronal mass ejection, which is characterized by an outer loop, a dark region and a filament.

**Ionosphere:** Upper layer of the Earth's atmosphere where the predominant process is the ionization of atoms and molecules by solar radiation.

**Lagrange point:** The Lagrangian points are the five positions in an orbital configuration where a small object affected only by gravity can theoretically be part of a constant-shape pattern with two larger objects (such as a satellite with respect to the Sun and Earth).

**Liability:** The amount of exposure.

**Magnetosphere:** Area around the Earth in which matter is ionized and the Earth's magnetic field is the predominant factor.

**Magnetotail:** The extension of the magnetosphere in the anti-sunward direction as a result of interaction with the solar wind.

**Magnetic declination:** Denoted an angle on the horizontal plane between the magnetic north and geomagnetic north poles.

**Magnetic reconnection:** A plasma process that converts magnetic energy to plasma kinetic energy accompanied by a change in the magnetic field topology. It allows a transfer of magnetic flux and plasma between separate magnetic flux regions.

**Mitigation:** Actions taken to reduce the impact of a hazard.

**Peril:** Insurance name for a natural phenomenon with the potential to cause loss or damage.

**Photosphere:** Lowermost layer of the Sun's atmosphere, from which the majority of the Sun's light is radiated into space. In other words, the photosphere is the visible part of the Sun.

**Plasma:** Gas, the atoms of which are wholly or partially ionized.

**Realistic disaster scenario:** Catastrophe scenario used for exposure management.

**Reliability:** The probability that a system will perform its intended functions without failure, within design parameters, under specific operating conditions, and for a specific period of time.

**Return period:** An event that has a $1/T$ frequency of occurring in any year is said to have a T-year return period.

**Ring current:** The main signature of a GMD formed due to the motion of trapped energetic electrons and ions injected earthward from the plasma sheet in the magnetotail.

**Risk:** An expected loss due to a particular hazard for a given area and a reference period.

**Risk transfer:** The process of moving risk from one party to another.

**Scenario:** A representation of a possible event based on scientific analysis or expert knowledge.

**Solar cosmic rays:** Particle radiation emitted constantly by the Sun in all directions with the velocity on the order of the 100,000 km/s.

**Solar cycle:** The approximately 11 year quasi-periodic variation in the sunspot number. The polarity pattern of the magnetic field reverses with each cycle.

**Solar flare:** A sudden release of energy in the solar atmosphere lasting minutes to hours, from which electromagnetic radiation and energetic charged particles are emitted.

**Solar wind:** Particle radiation emitted constantly by the Sun in all directions with the velocity ranging from less than 10 km/s to approximately 400–500 km/s.

**Solvency II:** A pan-European regulatory regime for insures.

**Space weather:** Phenomenon determined by the most varied interactions between the Sun, interplanetary space, and the Earth.

**Spinning Reserve:** Generation and responsive load that is on-line and can begins responding immediately.

**Substorm:** Space weather manifestations that occur due to the injection of energetic charged particles by an explosive energy release from the near-Earth magnetotail into the nightside magnetosphere.

**Sunspot:** The spots on the Sun's photosphere, which appear dark as the embedded strong magnetic fields push the hot plasma out, hence reducing the temperature compared to the one of the surrounding area.

**Thermosphere:** The outermost layer of the atmosphere between the magnetosphere and outer space, where temperature rises with increasing altitude.

**Van Allen belt:** Zone around the Earth in the manner of a belt, in which a particularly large number of particles of cosmic rays are trapped and held by the Earth's magnetic field.

# Index

**A**

ACE, 131, 181
aurora, 1, 186
autotransformer, 53, 87
awareness, 55, 74, 111, 122, 142, 155,
        184, 198

**B**

benchmark, 141, 156, 201
blackout, 83, 147, 208, 212
bow shock, 17
business interruption, 218

**C**

Carrington event, 18, 23, 186
Carrington-type event, 49, 159, 181
catastrophe, 177
catastrophe model, 213
        exposure, 215
        hazard, 213
        loss, 217
        vulnerability, 216
co-rotating interaction region, 20
complex image method, 44
contingency, 124, 151, 191
coronal holes, 11, 13
coronal mass ejection (CME), 9, 12, 20,
        129, 181
        detection time, 131
cost-benefit analysis, 132
critical factors, 183, 204, 212
critical infrastructure, 205
        protection, 159

**D**

damage, 218
damage ratio, 216
Daniel K. Inouye Solar Telescope
        (DKIST), 2
DC blocking device, 136, 196
disaster, 176
DSCOVR, 131

**E**

economic loss, 205
electricity market, 208
electrojet, 22, 44
European Space Agency (ESA), 129
exposure, 202
extreme-value model, 69

**F**

failure scenario, 82, 115, 151, 177, 207,
        213
FERC, 140, 143, 156
forecast, 128, 154, 157
        error, 133
        models, 132

**G**

Gauss, Carl Friedrich, 37
Geiger-Mueller counter, 129
geoelectric field, 42
geomagnetic activity index
        Dst, 20, 21, 187
        Kp, 19, 21
geomagnetic data, 37
geomagnetic disturbance, 18, 180, 185
        probability, 187
geomagnetic field, 17, 180
geomagnetically induced current (GIC),
        82, 182, 185
        effective, 96, 110
        measurement equipment, 65
        modeling, 189
        observation, 66
geomagnetically induced currents
        modeling, 49
Geomagnetism, 1
geomagnetosphere, 15, 19
grounding scheme, 196

**H**

Halloween event, 21, 25, 83
hazard, 176, 233
        map, 201

high harmonics, 91, 93, 114
Hydro-Québec event, 21, 24, 29, 85, 145, 186

**I**
insurance, 213
    claim, 186
interdependency, 207
INTERMAGNET, 38
interplanetary magnetic field, 16, 21, 185
ionosphere, 17
ionospheric auroral electrojet current model, 43
IZMIRAN, 29, 134

**L**
Lagrangian point
    L1, 130
    L5, 132
Lehtinen-Pirjola method, 51, 52

**M**
magnetosphere, 17, 129
magnetotelluric surveys, 47
mitigation actions, 126, 200
multi-hazard event, 177, 183

**N**
N-1 principle, 147
N-2 principle, 155
National Aeronautics and Space
    Administration (NASA), 129
natural hazard, 1, 115, 159, 176
NERC, 124, 140, 156, 212
Newton-Raphson solution, 111
nodal admittance matrix, 51

**P**
Parker Solar Probe, 130
photosphere, 8, 11
pipeline, 31, 53
plane-wave method, 42
power quality weak link, 197
power system
    architecture, 147, 188
    decentralized, 115
    modeling, 109, 144
    operation, 146
    operation state, 191

stability, 111, 147, 204
power system equipment, 188, 197
    circuit breaker, 107
    HVDC, 107
    measurement transformer, 102, 142
    power transformer, 85, 140
    series capacitor, 108, 136
    shunt reactor, 106
    synchronous machine, 97
    transmission line, 108
power transformer, 53, 85, 211
    construction, 86
    mitigation, 142
    saturation, 89
    single phase, 85, 87, 90, 101
    thermal assessment, 94, 141
    three phase, 85, 87, 90, 101
    vulnerability, 90, 96
    winding connection, 53, 88, 91

**R**
radiation belts, 16, 129
railroad storm, 23, 28
railway, 28
realistic disaster scenario, 217
recovery time, 85
relay protection, 104
reliability, 121
remedial action, 153
resiliency, 122, 160, 179, 234
    community, 199
    infrastructure, 125
    operational, 125
    socio-technical, 127
resiliency assessment, 126
ring current, 17, 19, 21
risk, 176, 213, 233
    assessment, 115, 122, 157
    awareness, 158
    emerging, 122, 180, 199
    governance, 233
    identification, 155
    management, 2, 126, 157
    matrix, 122, 159, 179
risk tolerability, 201
robustness, 55, 124, 125
    measurement transformer, 103
    power grid robustness to GMD, 3, 180, 184, 189, 203, 219, 234

power grid state, 109
power system equipment, 108, 197
power transformer, 91, 142
synchronous machine, 98

**S**
Slavgaard-Mansurov effect, 15
SOHO mission, 130
solar atmosphere, 8
solar cycle, 10, 21
solar wind, 8, 12, 16, 19, 130, 185
Solvency II, 217
Space Weather Operations, Research and
        Mitigation (SWORM), 156
Space Weather Prediction Centre
        (SWPC), 128
spinning mass, 196
STEREO mission, 129
storm onset, 21
storm sudden commencement, 21, 185
substorm, 18, 20, 185, 212

sudden impulse, 21, 185
Sun, 8, 129
sunspot, 9, 11, 21

**T**
total harmonic distortion, 93, 98
transmission system operator, 130, 147,
        151

**U**
UK Met Office, 128
United Nations Office for Disaster Risk
        Reduction (UNISDR), 176

**V**
Van Allen, James, 129
voltage avalanche, 149
von Humboldt, Alexander, 1, 37
vulnerability, 195, 200

**W**
worst-case scenario, 186